Physical properties of
liquid crystalline materials

Physical properties of liquid crystalline materials

W. H. DE JEU

GORDON AND BREACH SCIENCE PUBLISHERS
New York   London   Paris

Copyright © 1980 by Gordon and Breach, Science Publishers, Inc.

Gordon and Breach Science Publishers Ltd.
42 William IV Street
London WC2 4DE

Gordon and Breach, Science Publishers, Inc.
One Park Avenue
New York NY10016

Gordon & Breach
7-9 rue Emile Dubois
F-75014 Paris

Library of Congress Cataloging in Publication Data

Jeu, Wilhelmus Hendrikus de, 1943-
Physical properties of liquid crystalline materials.

Includes bibliographical references.
1. Liquid crystals.   I. Title.
QD923.J48   548'.9   79-3826
ISBN 0-677-04040-7

Contents

Preface

During the last decade the interest in liquid crystals has strongly increased
and one could say that at least the nematic phase is phenomenologically
well understood. The situation is different when it comes to the molecular
physics of liquid crystals. In many cases there is still little insight in the
molecular factors that determine the magnitude and anisotropy of a
physical property. This book is an attempt to review our knowledge in this
field. The choice of topics reflects strongly my own (practical) research
interest, and could easily have been different. Emphasis is mainly on the
nematic phase. Each of the chapters 3 to 7 discusses one topic, more or less
along the same lines. After introducing the specific property experimental
methods are considered, then typical results are discussed, and as far as
possible the relevant molecular factors are indicated. The main part of the
book is preceded by a general introduction and by a chapter on sample
preparation. In this chapter I have tried to collect much down-to-earth
information, that is nevertheless crucial for doing meaningfull experiments,
and cannot easily be obtained from the literature.

Theory as far as presented is very much a "poor experimentalist's
theory", emphasizing parts relevant to the experimental results and with
little consideration for general and/or elegant formalisms. I have tried to
present the material in a simple fashion, even at the risk of oversimplifying
complicated matters. For that reason I have always preferred, for example,
a simple geometrical argument above a general transformation rule. I trust
this will make the book readable for a larger audience. SI units are used
throughout the book. Although still not very common in the literature on
liquid crystals, its advantages are very clear—especially when both magnetic and electric properties are considered—in spite of the occurrence of
the factors ε_0 and μ_0^{-1}.[†] To allow easy comparison with the existing
literature in CGS units, in chapters 4 and 5 ε is not used for the
combination $\varepsilon_0\varepsilon_r$ (as officially recommended), but stands for the relative
permittivity ε_r, while ε_0 is always left explicit. The references are chosen
such that the reader who wants more details is probably helped best, thus
not necessarily doing justice to historical priorities. Many of the examples
are chosen from my own research, not so much because I think they are

---

[†] $\mu_0 = 4\pi \times 10^{-7}$ Hm^{-1} (exact), using the velocity of light in a vacuum $c = 2.998 \times 10^8$ ms^{-1},
one then calculates $\varepsilon_0 = 1/(\mu_0 c^2) = 8.854 \times 10^{-12}$ Fm^{-1}.

necessarily better than other ones, but because I happen to know them well, with all details readily available.

Much of the work on this book was done while the author was still at the Philips Research Laboratories in Eindhoven. I certainly owe much to the eight years of cooperation with my colleagues there: C. Z. van Doorn, C. J. Gerritsma, W. J. A. Goossens, W. van Haeringen, A. M. J. Spruijt, M. de Zwart, and especially J. van der Veen, from whom I learned so much about molecules and whose creative synthesis of new mesomorphic materials has been a constant source of inspiration. All of them have kindly given me their valuable criticism on early drafts of the various chapters. The same applies to A. J. Dekker and F. Leenhouts of the Solid State Physics Laboratory here in Groningen, to D. C. van Eck (Utrecht), to J. Nehring (Baden), and to W. Helfrich (Berlin). This book would not have been written if G. W. Gray (Hull) had not asked for it, and I want to thank him for his stimulating interest.

W. H. de Jeu
Groningen

January 1979

CHAPTER 1

Introduction

The *liquid crystalline* phase is a state of matter that is sometimes observed intermediate between a solid crystal and an isotropic liquid. A liquid crystal can flow like an ordinary liquid, while other properties, as for example the birefringence, are reminiscent of those of a crystalline phase. This combination led to the name liquid crystal, which is in fact a contradiction in terms. Other words in use are *mesophase* (meaning intermediate phase) or *mesomorphic phase*.

To understand the nature of liquid crystals we recall what happens at the melting point of a crystal. In the crystal the centres of mass of the molecules are located on a three-dimensional periodic lattice. As a consequence X-ray diffraction patterns can be obtained that show sharp Bragg reflections. In the liquid state the three-dimensional translational symmetry has disappeared; there only remains some short-range order between the centres of mass of the molecules. The X-ray diffraction patterns show only diffuse reflections, corresponding to this short-range structure. In the case of anisotropic molecules, the crystalline state will not only possess positional order, but may also have long-range order of the orientations of the molecules. Although both types of order usually disappear at the melting point, this is not necessarily so. Liquid crystals are characterized by the fact that the positional order has disappeared (or is strongly reduced), while some degree of orientational order is still present. Usually the molecules are greatly elongated, and their long axes, except for the influence of thermal fluctuations, are parallel to each other.

When a substance showing a liquid crystalline phase is heated, at the melting point the solid changes into a rather turbid liquid. When observed between crossed polarizers, this fluid phase is found to be strongly birefringent. Upon further heating a second transition point is reached where the turbid liquid becomes isotropic and consequently optically clear (clearing point, see Figure 1.1). The melting point and the clearing point define the temperature range in which the mesophase is thermodynamically stable. Both phase transitions are first-order, as indicated by the occurrence of a latent heat and a jumpwise change in the density. At the clearing point, however, these quantities are usually an order of magnitude smaller than those at the melting point. A compound that possesses such a mesophase is called a mesogenic compound. We shall use the term liquid crystal for a

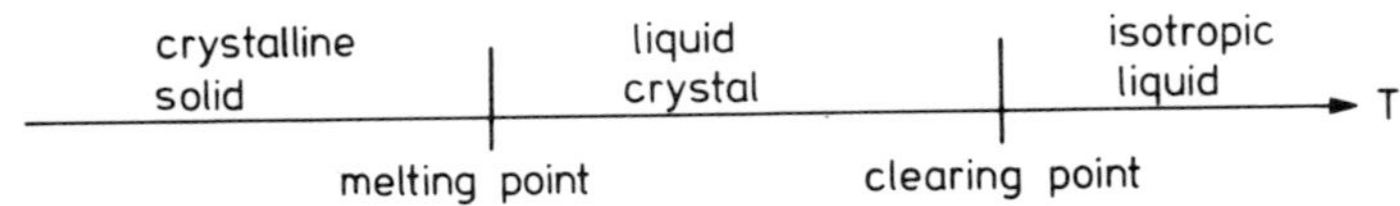

FIGURE 1.1   The liquid crystalline phase on the temperature scale.

mesogenic compound in its liquid crystalline phase. Liquid crystals of this type are called *thermotropic*. They can occur as single-component systems that show mesomorphic behaviour in a definite temperature range. Liquid crystalline phases can also exist intermediate between the solid state and an isotropic solution. The amount of solvent is then the most important variable: these are called *lyotropic* liquid crystals. We shall not discuss them explicitly in this book. Some general references to liquid crystals are given at the beginning of the reference list on page 124. An interesting account of the history of liquid crystals has been given by Kelker (1973).

Liquid crystals can be classified as *nematic* and *smectic* liquid crystals. In Table 1.1 we give some examples of nematics. The transition temperatures are given in a linear notation, where K stands for the crystalline state, N for the nematic phase and I for the isotropic liquid. The temperatures are in degrees Centigrade. The first compound, p, p'-dimethoxyazoxybenzene or p-azoxyanisole (PAA) is the "fruit fly" of liquid crystal research.

TABLE 1.1
Some examples of nematic liquid crystals

---

p-azoxyanisole, PAA (Arnold, 1964)

CH_3O—◯—N=N—◯—OCH_3 (with O on azoxy)   K 118 N 135.5 I

N-(p-methoxybenzylidene)-p'-butylaniline, MBBA (Kelker *et al.* 1970)

CH_3O—◯—CH=N—◯—C_4H_9 K 22 N 47 I

p-hexyloxy-o-hydroxy-p'-butylazobenzene (Van der Veen and Hegge, 1974)

$C_6H_{13}O$—◯—N=N—◯—C_4H_9 (with O—H)   K 8 N 82 I

p-heptyl-p'-cyanobiphenyl (Gray, 1975)

C_7H_{15}—◯—◯—CN K 28.5 N 42 I

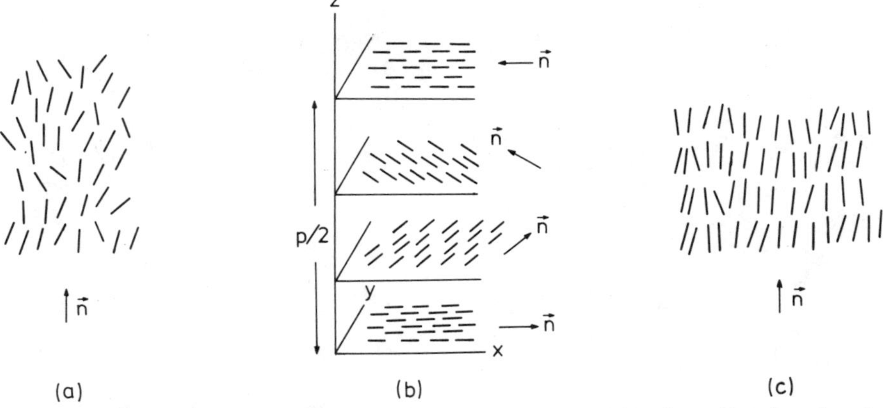

FIGURE 1.2 Nematic (a), chiral nematic (b) and smectic A (c) phase. For the sake of clarity the chiral nematic phase is pictured for the case of perfect alignment.

The melting point and nematic range are typical of many of the long-known mesogenic compounds. The other examples are chosen from more recently synthesized materials that have their nematic range around or slightly above room temperature. The essential features of the *nematic* phase can be summarized as follows (see Figure 1.2a).

a) The molecules are, on average, aligned with their long axes parallel to each other. Macroscopically a preferred direction is thus defined. Around this direction there is rotational symmetry: the phase is uniaxial.

b) There is no long-range correlation between the centres of mass of the molecules, which can translate freely. This property determines the fluid character of the nematic phase and results, for example, in nuclear magnetic resonance spectra of nematic liquid crystals consisting of narrow lines.

c) The axis of uniaxial symmetry has no polarity. Although the constituent molecules may be polar, this does not lead to a macroscopic effect.

It is convenient to describe the local direction of alignment by a unit vector n, the *director*, which gives at each point in a sample the direction of the preferred axis. Property (c) means that n and $-n$ are equivalent; the sign of n is physically of no importance. In an actual sample the orientation of n is imposed by the boundary conditions and possibly also by external fields. In Section 2.2 it is shown how these means are used to obtain a uniform director pattern. Without special measures the boundary conditions will vary over a substrate. This leads to variations in the director pattern, and thus to differences in birefringence. With a thin layer of a nematic between crossed polarizers under a microscope this is seen as a characteristic pattern or texture. The word nematic (Greek: νῆμα =

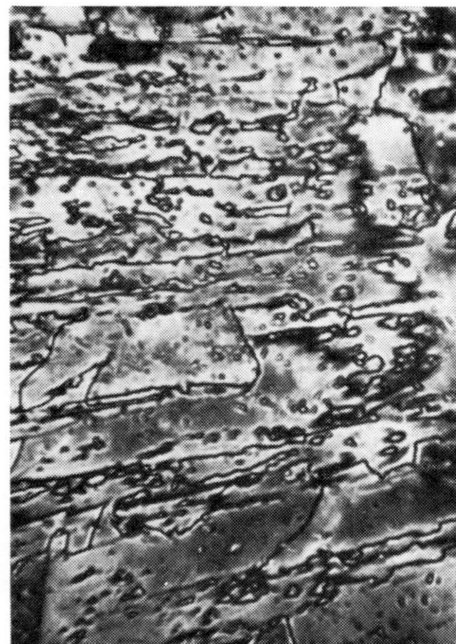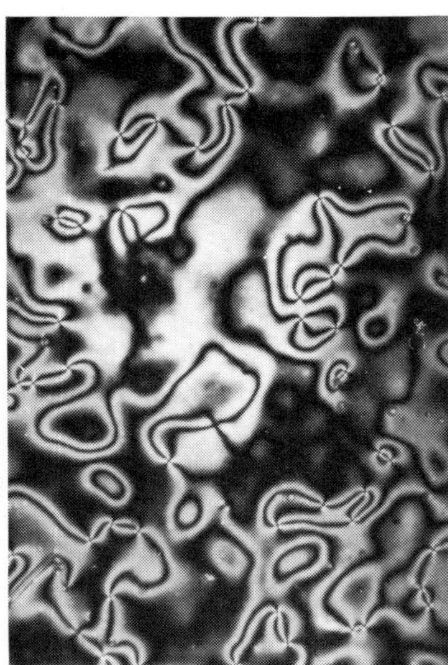

FIGURE 1.3   Textures of a thin nematic layer (crossed polarizers); (a) threaded texture, (b) schlieren texture.

thread) reflects the threads often observed in these textures (see Figure 1.3a). The threads are discontinuities in a director pattern that in general varies continuously over distances much greater than the molecular dimensions. Figure 1.3b shows a schlieren texture, in which the threads (from which the dark brushes originate) are perpendicular to the layer.

A nematic phase can occur either with molecules that have mirror symmetry, or in the case of optically active or enantiomorphic molecules with a racemic mixture of right- and left-handed species. If in the latter case one type of molecules is in the majority a *chiral nematic* mesophase results (abbreviated as N*). A simple picture of this phase is sketched in Figure 1.2b. In addition to the long-range orientational order there exists a spatial variation of the director leading to a helical structure. If we consider a series of planes perpendicular to the helix axis, there is in each plane orientational order as in the nematic phase. However, the local direction of alignment of the molecules is slightly rotated in adjacent planes. The helical structure can be described by an intrinsic non-constant director (see Figure 1.2b):

$$n_x = \cos\alpha,$$
$$n_y = \sin\alpha, \qquad (1.1)$$
$$n_z = 0.$$

The helix axis has been taken parallel to the z axis with $\alpha = q_0 z + \text{constant}$. A full rotation of $\boldsymbol{n}$ is completed through a distance $p = 2\pi/|q_0|$, called the pitch. The sign of q_0 distinguishes between left-handed and right-handed chiral nematics. Because **n** and $-$**n** are equivalent the repetition period is $p/2$. In practice, p can vary between say 200 nm and infinity. Hence p is much larger than the molecular dimensions. This means that for adjacent molecules, the difference in preferred direction is extremely small: the local order is essentially the same as in nematics. In Table 1.2 we give some examples of chiral nematic liquid crystals. Many derivatives of cholesterol belong to this class. For that reason the word *cholesteric* is frequently used as an alternative to chiral nematic. Comparing the nematic and the chiral nematic phase the following observations are important:

a) A particular compound shows as a function of temperature never a phase transition from a nematic to a chiral nematic phase.

b) Two different compounds, one with a nematic and the other with a chiral nematic phase, can possess at lower temperatures the same additional smectic phase.

c) Both phases are completely miscible. Addition of a small amount of an optically active compound (whether itself mesomorphic or not) changes a nematic phase into a long-pitch chiral nematic.

d) The X-ray diffraction patterns of nematic and chiral nematic phases are very similar. In both cases there is no long-range positional order that would lead to sharp reflections.

e) A mixture of two compounds of opposite chirality can give a nematic phase; at a certain composition the pitch becomes infinite. At this compensation point, there are no anomalies in any other physical properties that would indicate a phase transition.

TABLE 1.2
Some examples of chiral nematic liquid crystals

N-(p-ethoxybenzylidene)-p'-(β-methylbutyl)aniline (Dolphin *et al.* 1973)

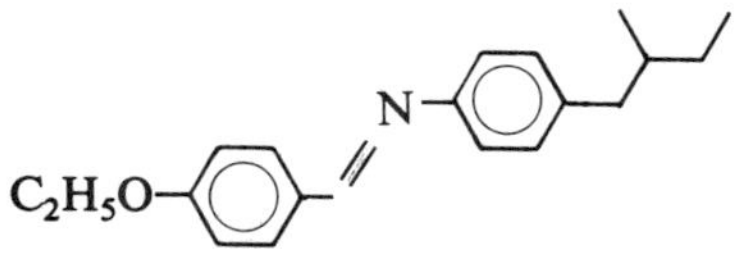

C_2H_5O—⬡—N=⬡— K 15 N* 60 I

cholesteryl chloride[a] (Leder, 1971)

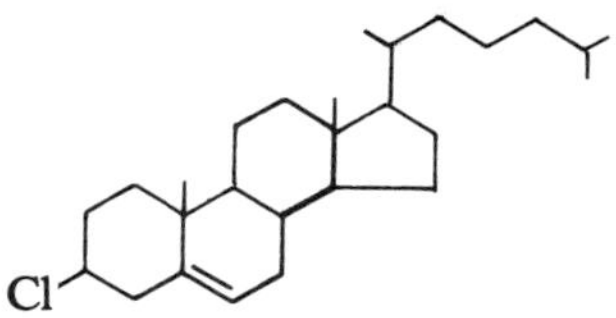

Cl— K 96 I (64 N*)

[a]This compound has a monotropic mesophase, which is only observed during supercooling of the isotropic liquid. Monotropic transition temperatures and phases are placed between parentheses.

From these results one must conclude that the chiral nematic phase is just a nematic phase that takes a somewhat different form because of the chirality of the constituent molecules. Alternatively one can consider the nematic phase as a chiral nematic phase with infinite pitch. Because of this similarity the word chiral nematic should be preferred to cholesteric. Due to the helical structure, the chiral nematic phase has unique optical properties (see, for example, Elser and Ennulat, 1967).

There are various kinds of mesophases that are called *smectic*. In general they are much more viscous than the nematic phase (Greek: $\sigma\mu\tilde{\eta}\gamma\mu\alpha=$ soap). They have the following properties:

a) The elongated molecules exhibit orientational order as in the nematic phase.

b) In addition, there is some positional order. The centres of the molecules are, on average, arranged in equidistant planes.

Direct evidence of the layered structure is obtained from the X-ray diffraction pattern, where a sharp reflection is observed at small Bragg angles, which corresponds to the layer thickness. The simplest type of smectic phase (smectic A or S_A) is depicted in Figure 1.2c. In fact several types of smectic phases have been observed. The main ones can be constructed in the following ways:

— The director can be parallel to the layer normal or tilted. If there is long-range order in the tilt directions the smectic phase becomes biaxial.

— Additional positional ordering of the molecular centres may occur within the smectic layers.

The various phases that can be obtained in this way are given in Table 1.3. For the corresponding X-ray diffraction patterns see, for example, de Vries (1975). A discussion and photographs of the various textures can be found in the book by Demus and Richter (1978). In the case of enantiomorphic molecules, the tilted phases give a chiral smectic phase (S_{C*}, *etc.*) in which

Table 1.3
Various types of smectic phase

	Director parallel to layer normal	Director tilted away from layer normal
Liquid layers	A	C
Pseudo-hexagonal packing in layers	B	G
Herringbone-type packing in layers	E	H

the tilt directions of adjacent layers are slightly rotated. Table 1.4 gives some examples of smectic phases. As we see a mesogenic compound can exhibit a considerable number of mesophases. This phenomenon is called polymorphism. In contrast to the NI or SI transition, the latent heat associated with a SN or SS transition can be extremely small, and probably second-order transitions may occur. There is still some disagreement about the classification of the various smetic phases. Tables 1.3 and 1.4 are consistent with the nomenclature F used in Halle (Demus and Richter, 1978), but in the literature sometimes the symbols G and H are interchanged. In what follows we shall restrict ourselves mainly to the uniaxial phases, with emphasis on the nematic phase.

Recently, mesophases have been discovered which are formed by disc-like molecules (Chandrasekhar *et al.* 1977; Billard *et al.* 1978). In these phases the molecular planes are parallel. We shall not discuss these latter types of phase.

TABLE 1.4
Some examples of smectic liquid crystals

p,p'-dinonylazobenzene (de Jeu, 1977)	
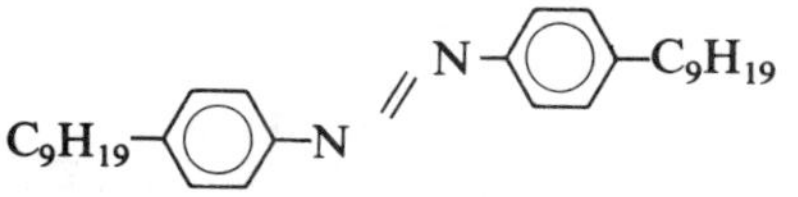	K 38 S_B 41 S_A 54 I
p,p'-diheptyloxyazoxybenzene, HOAB (Arnold, 1964)	
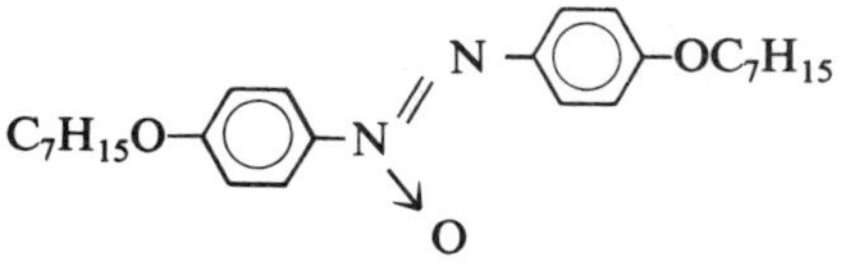	K 74.5 S_C 95.5 N 124 I
terephthal-*bis*-(*p*-butylaniline), TBBA (Doucet *et al.* 1974)	
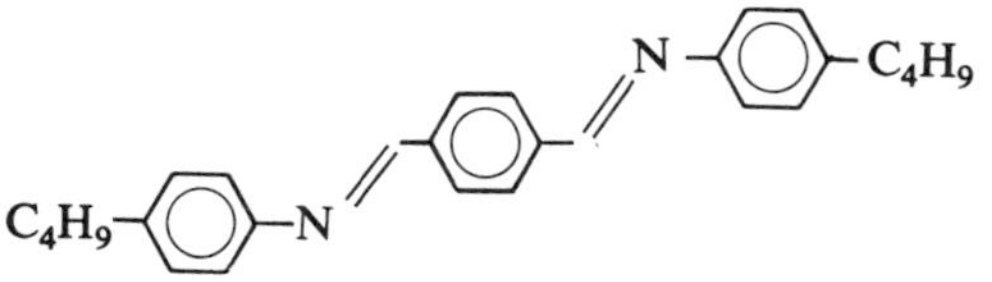	K 112.5 S_G 144 S_C 172.5 S_A 198.5 N 235.5 I (89.5 S_H 80 S_5)
cholesteryl myristate (Poziomek *et al.* 1972)	
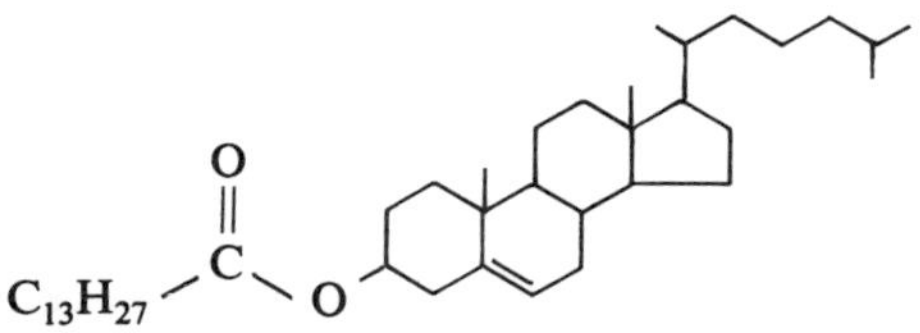	K 71 S_A 81 N* 86.5 I

Taking the nematic phase as an example, we consider now a sample with a uniform director pattern. We choose a laboratory-fixed coordinate system x, y, z such that the director is along the z axis. Hence the directions x and y are equivalent. The nematic phase has a lower symmetry (is "more ordered") than the isotropic liquid. To put this on a quantitative basis, we need to define an order parameter which is non-zero in the nematic phase and vanishes in the isotropic liquid. We shall discuss this concept from the molecular point of view, assuming for simplicity that the molecules are rigid. To specify the orientation of a molecule with respect to the laboratory system we introduce a molecule-fixed coordinate system ξ, η, ζ. The ζ axis is taken along the long molecular axis. The orientation of a molecule is completely determined by the three Euler angles θ, ψ and ϕ (see Figure 1.4; Margenau and Murphy, 1956, p. 286);

θ: angle between the z axis and the ζ axis;

ψ: angle between the ξ axis and the normal to the z, ζ plane, describing a rotation of the molecule around its long axis.

ϕ: angle between the x axis and the normal to the z, ζ plane, describing a rotation of the whole molecule around the director.

The third Euler angle ϕ can be disregarded because of the uniaxial symmetry of the nematic phase. The average orientation of the molecules can be specified by an orientational distribution function $w(\theta, \psi)$, where $w(\theta, \psi) \sin\theta \, d\theta \, d\psi$ is the fraction of molecules with Euler angles between θ and $\theta + d\theta$, and ψ and $\psi + d\psi$. The average value of a quantity X over the orientations of all molecules is then given by

$$\langle X \rangle = \int_0^\pi d\theta \int_0^{2\pi} d\psi \, X w(\theta, \psi) \sin\theta. \tag{1.2}$$

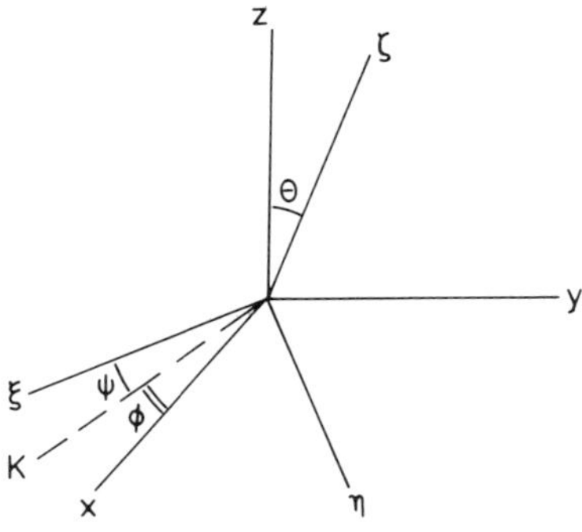

FIGURE 1.4   Euler angles defining the molecular frame ξ, η, ζ with respect to the laboratory frame x, y, z; K is the normal to the z, ζ plane.

From the properties of the nematic phase, the dependence of $w(\theta, \psi)$ on ψ is expected to be weak. On the other hand, the distribution function will be strongly peaked around the direction $\theta = 0$ or $\theta = \pi$ (the director), while for $\theta = \frac{1}{2}\pi$ we expect a minimum for $w(\theta, \psi)$. Because $\theta = 0$ and $\theta = \pi$ are equivalent, we have $w(\theta, \psi) = w(\pi - \theta, \psi)$. Consequently, if X in Eq. (1.2) changes sign when θ is replaced by $\pi - \theta$, its average value will be zero.

In practice one does not want to work with the full distribution function (which cannot be determined completely anyhow), but with order parameters that are defined in terms of the angles θ and ψ. Looking for an order parameter describing the average orientation of the long axis with respect to the director, we consider ζ_z, the component of a unit vector along the ζ axis on the director. The average value of $\zeta_z = \cos\theta$ is zero, however. The next choice is $\zeta_z^2 = \cos^2\theta$, which is used to define an order parameter S as

$$S = \tfrac{1}{2}\langle 3\zeta_z^2 - 1 \rangle$$

$$= \tfrac{1}{2}\langle 3\cos^2\theta - 1 \rangle. \tag{1.3}$$

If the distribution of the long molecular axes is random, as in the isotropic phase, we have $\langle \cos^2\theta \rangle = \frac{1}{3}$, and $S = 0$. The value $S = 1$ corresponds to the case of perfectly aligned molecules. In practice, S varies from values around 0.3–0.4 at T_{NI} to about 0.8 at much lower temperatures. To specify the average orientation of the molecules completely, we must now give in addition either $\xi_z^2 = \sin^2\theta \, \sin^2\psi$ or $\eta_z^2 = \sin^2\theta \cos^2\psi$. It is convenient to take instead (Alben *et al.* 1972)

$$D = \tfrac{3}{2}\langle \eta_z^2 - \xi_z^2 \rangle$$

$$= \tfrac{3}{2}\langle \sin^2\theta \cos 2\psi \rangle. \tag{1.4}$$

A finite D means that there is a difference in tendency of the two transverse molecular axes to project on the z axis. It does not mean that the nematic phase is biaxial. There is no preference for either the ξ axis or the η axis of different molecules to be parallel.

More generally one can ascribe to a molecule in a nematic phase an order parameter tensor $\mathbf{S}$ with elements S_{ij}, $i, j = \xi, \eta, \zeta$ (Saupe, 1963; see also Diehl and Khetrapal, 1969). The tensor $\mathbf{S}$ is real, symmetric and has zero trace; consequently there are five independent elements. In the above discussion we have simplified this situation by taking the molecular coordinate system such that $\mathbf{S}$ is diagonal. In practice this is always possible, and one is left with only two different elements for which we have chosen $S = S_{\zeta\zeta}$ and $D = S_{\eta\eta} - S_{\xi\xi}$.

In the case of real molecules the first problem is how to define the molecular frame. Usually the relatively rigid aromatic core is considered to be the most relevant part for this choice. For molecules with a *trans* configuration (Table 1.1, examples 1–3), the axis through the two outer *para* carbon atoms of the aromatic part is often taken as the ζ axis. If the various benzene rings are coplanar (as for example with azobenzenes), the ξ, ζ plane is identified with this plane. Where such a coplanarity is not present, a plane containing the ζ axis and making equal angles with the planes of the benzene rings can be chosen. In practice, the differences between the various choices that are possible are probably not very important. A second problem is that mesogenic molecules usually contain flexible groups such as, for example, alkyl chains. In principle, then, order parameters should be defined for each CC bond; these parameters will have different values because of the differences in flexibility. These effects can be observed with techniques such as nuclear magnetic resonance spectroscopy (see, for example, Emsley *et al.* 1975). A discussion of this is outside the scope of this book. We shall in general assume that, in spite of these considerations, a rigid body can be used to represent a molecule in its "average" conformation. In that context two models are often used to represent real molecules.

(1) *Rigid rods.* Rigid rods, as for example prolate spheroids, cylinders or spherocylinders (cylinders capped with hemispheres), are the simplest types of object which allow nematic behaviour. They have complete cylindrical symmetry around the ζ axis. Evidently this type of symmetry is not present in most mesogenic molecules. Nevertheless, the model is useful, because the molecules can often be assumed to rotate freely around the ζ axis. In that case, there is cylindrical symmetry and $w(\theta, \psi)$ is independent of ψ. Thus we have $D = 0$ and S is the only order parameter that occurs in a description of the nematic phase.

Maier and Saupe (1959, 1960) have given a molecular-statistical theory of the nematic phase with one order parameter. In a mean-field approximation each molecule is assumed to experience an average potential $W_i(\theta)$ given by

$$W_i(\theta) = -(A/V^2)S\left(\tfrac{3}{2}\cos^2\theta_i - \tfrac{1}{2}\right), \tag{1.5}$$

where A is the strength of the potential and V is the molar volume. The distribution function is related to the potential by $w(\theta) = (1/Z)\exp(-W_i/k_BT)$, where Z is the normalization constant. Maier and Saupe considered the interaction between the induced dipoles of the molecules as the basis of their theory. In that case, the interaction strength A is approximately proportional to the squared anisotropy of the molecu-

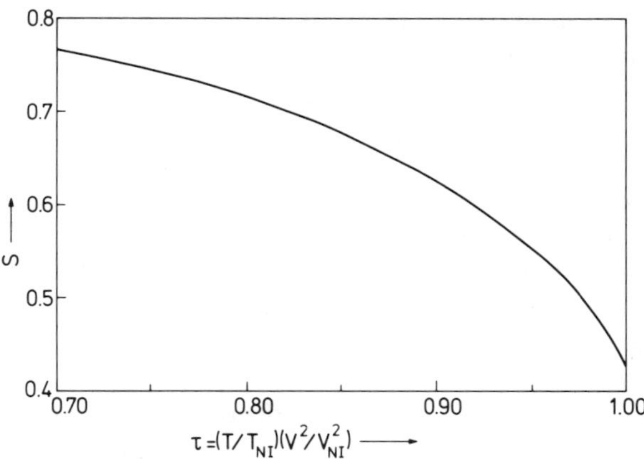

FIGURE 1.5   Plot of the order parameter S *versus* reduced temperature according to Maier and Saupe's theory.

lar polarizability. In practice, A is often considered as an empirical parameter that also takes other types of interaction (for example repulsions between the anisotropic molecular shapes) into account. Once $W_i(\theta)$ is known, the free energy difference between the isotropic phase ($S=0$) and the nematic phase (finite S) can be calculated. The order parameter is then predicted to be a universal function of the reduced temperature $\tau = TV^2/(T_{NI}V_{NI}^2)$, with $S=0.43$ at T_{NI} (see Figure 1.5). For the temperatures of interest this function can be approximated analytically to within 1% by

$$S = (1 - 0.98\tau)^{0.22}. \tag{1.6}$$

Although the predictions of Maier and Saupe's theory are certainly not correct quantitatively, for many liquid crystals they are found to be a good approximation.

(2) *Rigid bodies with two mirror planes.*   As a second model, we consider an anisotropic rigid body that is not axially symmetric, but still has two mirror planes, that are perpendicular to each other and both contain the ζ axis. We shall choose the ξ axis in one plane and the η axis perpendicular to it. One can think of a prolate ellipsoid with its shortest axis in the η direction, or of a rectangular parallelipiped. As in model (1) we can use S to specify the average degree of orientation of the long molecular axes. In addition D may have a finite value. As said before, D measures the difference in tendency of the two transverse molecular axes to project on the z axis. D can only be finite if $S \neq 1$, as otherwise such a projection is zero. On the other hand, in the other limit where $S=0$ (no long-range

orientational order of the ζ axes), D is also necessarily equal to zero. Hence D will show a behaviour like

$$D \sim S(S-1). \tag{1.7}$$

Luckhurst *et al.* (1975) have extended Maier and Saupe's theory of the nematic phase to the case of non-axially symmetric molecules. In fact the maximum of D is found at $S \approx 0.4$. As this is approximately the value of S at T_{NI}, we can, in practice, expect D to decrease with decreasing temperature. Of course all this is important only if the molecules do not rotate freely around the ζ axis.

Finally we note that in discussing the degree of order, we have assumed so far that the director is uniform along the z axis. Unless special measures are taken this will not be true for a real liquid crystal. However, the variations of $\boldsymbol{n}$ are only important over distances much larger than the molecular dimensions, and we can still take a nematic to be locally uniaxial. S then specifies the average degree of orientation with respect to the local director. In other words, the space dependence of $\boldsymbol{n}$ and the temperature dependence of S can be treated separately. This property is the basis of the continuum theory of liquid crystals, which will be discussed in Chapters 6 and 7. Up to that point, we can think of the liquid crystal as having a uniform director pattern.

CHAPTER 2

Sample preparation

2.1 CHEMICAL PROPERTIES

To study the physical properties of liquid crystals it is of the utmost importance to have clearly defined transition temperatures. In general these temperatures will depend on the purity of the compound. The classical criterion for purity is the value and the sharpness of the melting point, as observed, for example, on a small amount of substance under a polarizing microscope equipped with a heating stage. The higher the purity of a particular compound, the higher the melting point and the narrower the temperature range over which the melting takes place. This is due to the fact that addition of another substance usually reduces the melting point and broadens the melting range (illustrated in Figure 2.1 for a binary mixture). In principle a substance should be recrystallized several times until no further change in the melting point occurs.

In the case of liquid crystals the clearing point provides an additional criterion for purity. In a binary mixture, the clearing temperature usually varies roughly linearly with composition (see Figure 2.1). In a similar way the presence of a non-mesogenic compound leads to a decrease of T_c. Hence the higher T_c, the better is the purity. Exceptions may occur in the rare cases where the impurities themselves are mesomorphic. In general, the clearing range of a substance should be smaller than 0.2°C at heating rates lower than 1°C/min. Measurements of the transition heats by calorimetric methods can provide additional information on the purity of a compound. For special purposes, the above-mentioned requirements may not suffice. For example, traces of ferromagnetic impurities that hardly affect T_c may have an appreciable influence on the diamagnetic susceptibility; dissociating or ionic impurities are of special importance in studies of the electrical conductivity.

The synthesis of a mesogenic compound is in principle not different from that of a related compound that does not form a mesophase. However, the extra demands on purity usually require a high degree of specificity of the reactions used, so that no (non-mesomorphic) side products are formed. Often it will be necessary to start with highly purified basic materials. A lucid discussion of the synthesis of liquid crystals for physicists has been given by Keller and Liebert (1978). For the purposes of

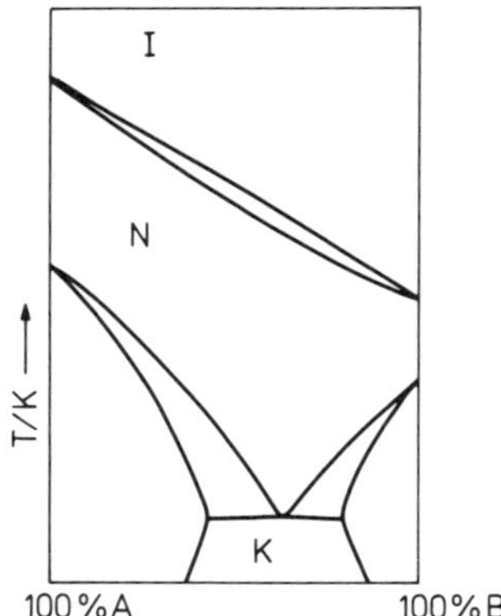

FIGURE 2.1 Phase diagram for a hypothetical binary mixture of two mesogenic compounds A and B.

this book, we shall restrict further discussion to some general points that are also of importance to the non-chemist who handles liquid crystals.

Most of the classes of liquid crystals mentioned so far are chemically fairly stable at atmospheric pressure and at temperatures within their mesomorphic range. Schiff's bases are the most noticeable exception. These compounds are synthesized by the condensation of say a *para*-substituted benzaldehyde and a *para*-substituted aniline:

As this is an equilibrium reaction, Schiff's bases are sensitive to even small amounts of water (Sorkin and Denny, 1973), and once the starting products are present, the aniline is subject to oxidation. For this reason it is useful to degas Schiff's bases in a vacuum oven. In practice the hydrolysis can be quite serious. Pure MBBA (R_1 is CH_3O, R_2 is C_4H_9) has $T_{NI} =$ 47°C, but without special precautions this temperature will drop readily to values lower than 40°C. Simply by storing MBBA over molecular sieves, a clearing point of the order of 45°C can be maintained. Often one has to be satisfied with this value, as it is very difficult to perform experiments under conditions that allow a higher clearing point to be maintained over a long period. A related effect can occur in mixtures of Schiff's bases. In a binary mixture of the compounds with interchanged end groups will be formed after some time (combinations R_1/R_4 and R_3/R_2). Thus one ends up with a quaternary mixture that can have properties quite different from those of the original binary mixture. This effect can be circumvented using components of binary mixtures which always have either $R_1 = R_3$ or $R_2 = R_4$.

Phenyl benzoates (esters) can also be subject to hydrolysis:

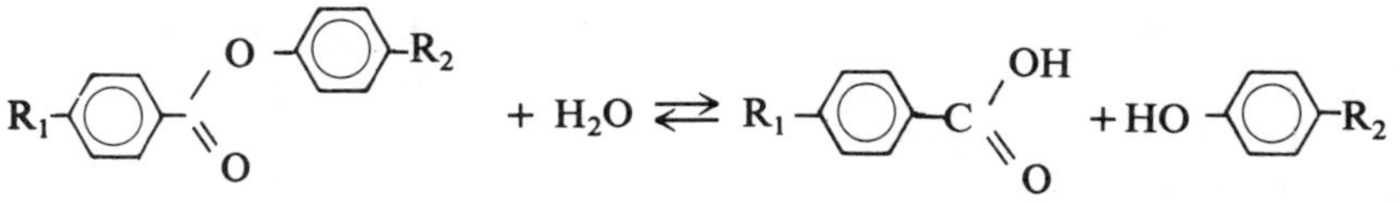

This effect is much less important than in the case of Schiff's bases. Nevertheless, problems can arise, because the resulting carboxylic acid can act as a surfactant, causing a homeotropic orientation of the director (see Section 2.2). This effect can be important even at very small concentrations of the order of 0.1%.

Under the influence of light (especially in the UV region) azobenzenes are subject to *trans-cis* isomerization:

trans   *cis*

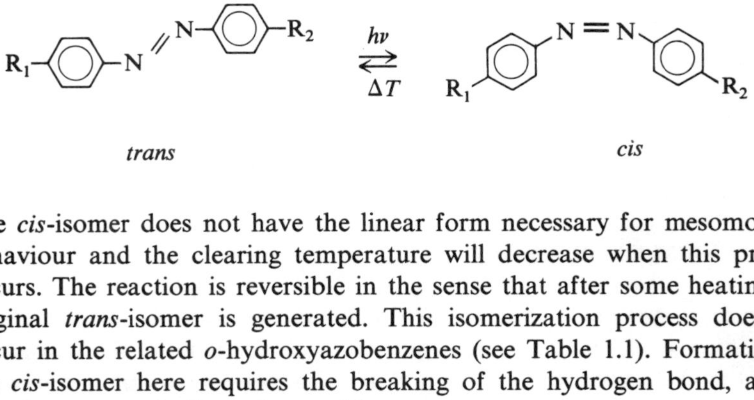

The *cis*-isomer does not have the linear form necessary for mesomorphic behaviour and the clearing temperature will decrease when this process occurs. The reaction is reversible in the sense that after some heating the original *trans*-isomer is generated. This isomerization process does not occur in the related *o*-hydroxyazobenzenes (see Table 1.1). Formation of the *cis*-isomer here requires the breaking of the hydrogen bond, and is therefore energetically unfavourable.

Trans-cis isomerization can also occur in azoxybenzenes (Suh-Bong Rhee and Jaffé, 1973). Experiments with p, p'-dibutylazoxybenzene ($T_{NI} = 32°C$) indicated that T_{NI} decreases strongly upon irradiation with UV light of 365 nm. In addition, the liquid turned *deep* yellow owing to the different light absorption of the *cis*-isomer. After heating for about 10 minutes at 80°C, the original *light* yellow colour returns while T_{NI} is again found at 32°C (sharp). An alternative process involving the photo-Wallach rearrangement

would not have been reversible in this way. This latter reaction is probably important only if the medium has a relatively high dielectric permittivity.

Stilbenes are also subject to photochemical *trans-cis* isomerization (Sargent and Timmons, 1964). For this class of compounds, the *cis*-isomer can in fact be quite stable. In the presence of oxygen, the *cis*-isomer may undergo further photochemical cyclodehydrogenation to a phenanthrene:

For the various substituents X that have been used, the photochemical stability decreases in the order $H > Cl > CN$.

2.2   UNIFORM SAMPLES

The uniaxial symmetry around the director in the nematic phase leads to an anisotropy in many physical properties. For example, the refractive index, the dielectric permittivity and the magnetic susceptibility have a different value parallel to the director ($\parallel$) and perpendicular to it ($\perp$). In order to measure these quantities, samples with a well-defined, uniform director pattern are required. A pattern of this kind can be obtained in thin layers (typically 5–200 μm), through interaction of the liquid with the glass walls of the sample. Two limiting cases are important: *homeotropic* layers, in which the director is everywhere perpendicular to the walls, and *uniform planar* layers, in which the director is parallel to one direction in the plane of the substrate (see Figures 2.2a and 2.2b). Of course, intermediate situa-

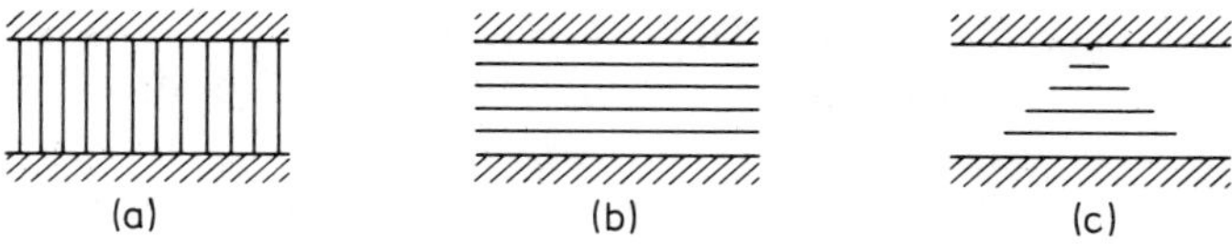

FIGURE 2.2 Homeotropic (a), uniform planar (b), and twisted planar (c) director pattern.

tions involving a uniform director pattern with tilt angles between 0° and 90° can also be obtained. Alternatively, external fields may be used to obtain a uniform director pattern. We shall first consider, however, surface-produced alignment. For reviews of this subject see Kahn *et al.* (1973), and Guyon and Urbach (1976).

The interaction of a mesogenic compound with a substrate will depend on the nature of both the compound and surface. We shall restrict ourselves here mainly to glass surfaces and shall not differentiate between various types of compound. Consequently, not all the methods mentioned can be expected to give the same results in all cases. At the surface of glass, depending on the sample's history, mainly siloxan (I) and silanol (II) groups will be present (see, for example, Boehm, 1966):

An SiO_2 substrate treated at elevated temperatures (800–1000°C) will be mainly in the form (I). The siloxan groups are opened by a treatment with alkali hydroxides, leading to silicate groups. On subsequent exchange of alkali by hydrogen by application of an acid the silanol groups are formed. Mesogenic molecules with polar end groups will then orientate themselves normal to the surface, the coupling being due to hydrogen bonds or dipole-dipole interaction. Once a mono-layer is coupled to the substrate in this way, its average orientation is transmitted through the bulk *via* the orientation-dependent intermolecular forces that give rise to the mesophase itself, leading to a homeotropic texture. Alternatively, for formation of the first monolayer, one can use a specially chosen surfactant consisting of a polar group that interacts with the surface, and a long aliphatic chain that couples with the liquid crystal. Among the compounds frequently used are cetyltrimethylammonium bromide (CTAB):

$$Br^{(-)}N^{(+)} - (CH_3)_3$$
$$|$$
$$C_{16}H_{33}$$

and other compounds such as lecithin. Furthermore, carboxylic acids have been used. The surfactants can either be applied to the glass surface beforehand, or added to the liquid crystal in small concentrations ($\sim 0.1\%$).

Since physical absorption is a reversible process, the adsorbed monolayer is not necessarily permanent. The surfactant molecules compete for the surface sites with other adsorbable molecules, including impurities and the mesogenic molecules themselves. Some of these problems can be circumvented if the monolayer is coupled chemically to the surface, using for example long-chain alcohols (ROH). At relatively high temperatures these compounds react with the silanol groups (Ballard *et al.* 1961), leading to substrates like

$$
\begin{array}{ccc}
\text{OR} & & \text{OR} \\
| & & | \\
\text{Si} & & \text{Si} \\
/ | \backslash & & / | \backslash
\end{array}
$$

For alkyl chains R from C_{12} to C_{16} there is an optimal interaction with the mesogenic molecules (Cladis *et al.* 1971), often giving a stable homeotropic orientation. The surface energies are estimated to be of the order of $0.02\ \mathrm{Jm^{-2}}$. The *difference* in surface energy between the homeotropic and the planar orientation may be an order of magnitude smaller ($10^{-3}\ \mathrm{Jm^{-2}}$). Methoxysilanes of the form $RSi(OCH_3)_3$ (Kahn *et al.* 1973) and organometal complexes (Matsumoto *et al.* 1979) have also been used.

It should be emphasized that several aspects of the processes described above are not yet fully understood. For example, the surface density of the agent that couples to the substrate is an important factor. Again, at low concentrations, CTAB causes a planar alignment because the long alkyl chains do not hang away from the substrate, but are parallel to it (Proust and Ter-Minassian-Saraga, 1975). The surface density in turn depends strongly on the experimental conditions. In this sense, one can still speak of the "art" of obtaining a specific boundary condition. The cleaning of the glass is also an essential step that should not be underestimated. Usually strongly oxidizing conditions are necessary to remove the last few layers of organic contaminants.

To obtain uniform planar samples, two steps are involved. First the boundary conditions have to be tangential to the surface, but this can lead to a continuously degenerate planar sample. A *preferred* tangential direction must be created to obtain a *uniform* director pattern. A planar sample can be made using surfactants with an alkylene chain and at least two polar groups, situated such that adsorption of the polar groups at the surface causes the alkylene chains to lie parallel to the substrate. By way of example we quote:

$$R\text{--}(CH_2)_n\text{--}R, \quad R \text{ is COOH, CN, NH}_2, \quad n = 3 - 11.$$

When alternatively a cleaning process using sodium hydroxide is used for the glass, good results can be obtained with polyoxyethylene alkyl ethers (sold as cremophor O by BASF, Ludwigshafen):

$$CH_3(CH_2)_n(OCH_2CH_2)_mOH, \quad n\approx15, \, m\approx20.$$

Excellent planar samples can be obtained by deposition of a polymer film on the substrate. Examples are polyvinylalcohol

$$[-CH_2-CHOH-]_n,$$

which can simply be deposited from an aqueous solution, and p-xylylene. The latter compound is obtained via vacuum pyrolysis of di-p-xylylene at temperatures of 600–700°C, and subsequent condensation and spontaneous polymerization of this reactive substance to form a linear polymer (trade name: parylene N) on a substrate at temperatures lower than 70°C (Gorham, 1966):

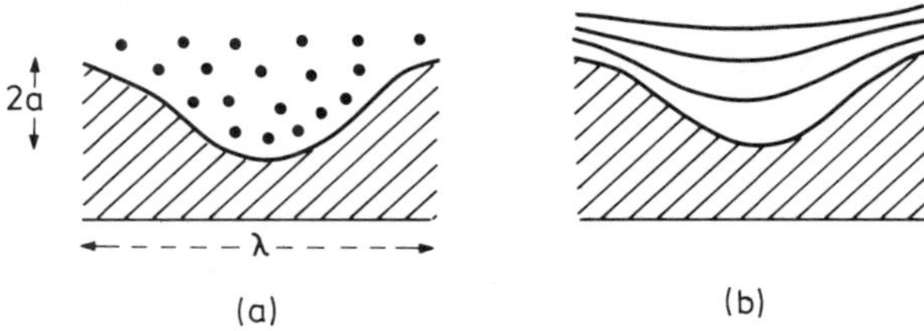

The process results in a non-polar surface layer with the aromatic rings parallel to the substrate. This ensures an optimal interaction with the aromatic rings of the adjacent layer of mesogenic molecules.

Once the boundary conditions are tangential, a preferred direction on the substrate can be obtained by rubbing the surface in one direction with a soft cloth or paper. This leads to scratches that give a surface topography with a surface energy that depends on the orientation of the director. The two extreme situations are sketched in Figure 2.3. The second case involves an elastic deformation of the director pattern and consequently has a higher energy. For a sinusoidal surface the difference in free energy

FIGURE 2.3 Director pattern parallel to a grooved surface (a) and perpendicular to it (b). The latter situation introduces a distortion of the director pattern.

between the two configurations has been calculated to be (Berreman, 1972)

$$\Delta F = (\pi^3 K / 2\lambda)(2a/\lambda)^2, \tag{2.1}$$

where K is an average elastic constant (see Chapter 6) and a and λ are defined in Figure 2.3. With $K \approx 10^{-11}$ N, $2a/\lambda \approx 5$ and $\lambda \approx 0.05$ μm one finds $\Delta F \approx 0.5 \times 10^{-3}$ Jm^{-2}, the same order of magnitude as the difference in physico-chemical coupling energies for a homeotropic and a continuously degenerate planar orientation. In practice, exact tangential boundary conditions are seldom obtained. Depending on the rubbing process, the director will be tilted away a few degrees from the substrate. Two substrates with the rubbing directions parallel can then be combined in two different ways, leading to director patterns in the layer as indicated in Figure 2.4. If the rubbing directions are perpendicular to each other a *twisted planar* sample is obtained (Figure 2.2c).

Efforts have obviously been made to find better defined methods than rubbing to obtain an appropriate surface topography. Such a method is to evaporate SiO, MgF$_2$, Bi$_2$O$_3$, Au or some other substance onto the glass, such that the molecules of the evaporated species are incident at a grazing angle to the substrate. Electron microscope photographs show the existence of anisotropic, elongated islands of the evaporated substance with their long axes in the same direction. Many liquid crystals give a uniform alignment on these surfaces. Depending on the evaporation conditions, especially the angle of incidence, and the liquid crystal considered, all kinds of tilt angles between 0° and 60° can be obtained (see, for example, Guyon and Urbach, 1976). This behaviour has been thought to result from a surface textured like the teeth of a saw. Tilt angles between 50° and 90° can be obtained by combining skew evaporation techniques with surfactants.

Alternatively to surface-produced alignment, one can apply an external magnetic field to obtain a uniform director pattern. As will be discussed in more detail in Chapter 3, the anisotropy of the magnetic susceptibility causes a difference in the free energy density of the liquid crystal when the field is parallel or perpendicular to the director. This difference is equal to $\frac{1}{2}\mu_0^{-1}(\chi_\parallel - \chi_\perp)B^2$, where B is the magnetic induction, μ_0 the permeability

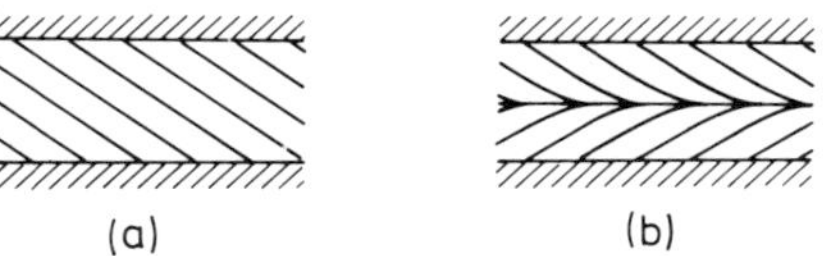

FIGURE 2.4 Different director patterns in a nematic layer that may be obtained with parallel rubbing directions of the enclosing glass plates and a small tilt angle at the substrates.

of vacuum, and χ the volume susceptibility. Since $\chi_\parallel$ is usually greater than $\chi_\perp$, the director thus tends to be parallel to the field. It should be emphasized that the field hardly affects the orientation fluctuations of the individual molecules. Any effect of this type is very small, like the Cotton-Mouton or Kerr effect in an isotropic liquid, and can be left out of account. It is the cooperative effect of the susceptibility anisotropy of a volume element that makes the director capable of being aligned. Typically $\chi_\parallel - \chi_\perp \approx 10^{-6}$ SI units, and for a magnetic induction of 1 T this leads to a free energy density of the order of 0.3 Jm^{-3}. In practice, there will be competition between the surface-induced alignment (intentionally or otherwise) and the effect of the field on the director. If the anchoring energy of the liquid crystal at the substrate is large compared with the magnetic energy (strong anchoring), there will be a boundary layer where the director is hardly influenced by the field. This layer may extend to a few microns (see Figure 2.5). For weak anchoring, it will be less. Often, strong anchoring is assumed, because this makes it possible to consider only the free energy of the bulk of the liquid crystal in combination with fixed boundary conditions. In practice it seems that this assumption is true only in some special cases. In homeotropic samples particularly there seems to be hardly any strong anchoring.

The competition between an external magnetic field and the boundary conditions can be used to measure the tilt bias angle. A possible method is as follows. A uniformly oriented nematic layer, as illustrated in Figure 2.4a, is rotated in a magnetic field until the orientation is found where a measured physical property is independent of the strength of the magnetic field. At this null orientation the tilt bias angle is equal to the angle between the magnetic field direction and the substrate. In practice, substrates coated with electrodes are often used, and the change in capacitance (due to a difference between $\varepsilon_\parallel$ and $\varepsilon_\perp$) is monitored. Alternatively, the optical retardation can be measured. A detailed discussion is given by Scheffer and Nehring (1977). The method does not work, of course, in the situation of Figure 2.4b. So far we have considered only external magnetic

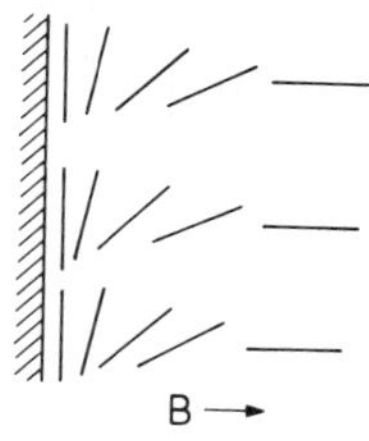

FIGURE 2.5 Boundary layer when a strong magnetic field is applied normal to a surface with uniform tangential boundary conditions.

fields. There are several reasons that make electric fields less suitable as a means of obtaining a uniform pattern. Because of the anisotropic conductivity (due to residual impurities) space charge will be generated and may cause hydrodynamic flow. Secondly, because the dielectric permittivity is relatively large, the electric field in a sample with a deformed director pattern will not be uniform.

Finally we consider a few other aspects of sample preparation. Often a thin nematic layer between glass plates will be required. These plates can be kept apart at a fixed distance by thin sheets of mylar, polyethylene or metal, or by wires of tungsten or nylon. The two glass plates, pretreated to obtain appropriate boundary conditions, can then be glued together with epoxy resin, or sealed together with a glass frit seal. When polymeric materials are used, it is necessary, before filling, to store the cell for about an hour at the highest temperature to be used, because these materials often undergo non-reversible thermal expansion. Preferably a vacuum oven should be used, to prevent volatile materials from the epoxy resins interacting with the glass surfaces. A thin cell is easily filled with the liquid crystal by capillary action. This should always be done at temperature above T_{NI} to prevent flow alignment of the director (see Chapter 7), which may influence the boundary conditions if the filling is done below T_{NI}.

If the layers are relatively thin (up to about 100 μm) the spacing of the empty cell can be measured accurately by interferometric methods. The sample is illuminated, and the light reflected from the two inner glass-air interfaces is analysed in a monochromator. The intensity will show a number of maxima and minima as a function of the wavelength λ. At a minimum, the optical path difference $2d/\lambda$, where d is the spacing of the cell, equals an integer, say k. Hence, if the first minimum of the run is at k_0 we have successive minima at wavelengths λ_l for which

$$2d/\lambda_l = k_0 + l, \quad l \text{ integer.} \tag{2.2}$$

Hence the slope of a curve of λ_l^{-1} *versus* l gives the thickness. For thicker samples, a microscope can be used to focus on the two inner interfaces, and the thickness can be read directly from the difference in the calibrated micrometer settings. In general, it is acceptable to take the thickness of the nematic layer after filling as being equal to the spacing of the empty cell. If desired this can be checked by filling a cell up to only a half, or by deliberately introducing an air bubble, by which means it is possible to measure the thickness very close to the nematic layer.

We conclude this chapter with some recipes that have been found to give well-defined boundary conditions for many liquid crystals, without being unique or universal. To obtain planar alignment we do not consider the (better) methods that require more elaborate equipment.

General procedure for the cleaning of a glass substrate

1) Five minutes ultrasonic cleaning in a 1% soap solution in water.

2) One hour in a 1M solution of KOH at 50°C.

3) One hour in an acid mixture at 60°C. (The acid mixture consists of one part concentrated HNO_3, one part concentrated H_2SO_4 and two parts water.)
After each step, rinse thoroughly with distilled water.

4) Rinse with *iso*-propanol.

Procedure for normal boundary conditions

If necessary reclean the glass plates as indicated above.

1) Dry the substrate in hot *iso*-propanol vapor.

2) Place the substrate for 30 minutes in dodedecan-1-ol at 200°C.

3) Degrease the substrate thoroughly in a vapour-condenser with *iso*-propanol for about 8 hours.
If necessary add 0.1% CTAB to the liquid crystal.

Procedure for uniform tangential boundary conditions

1) Prepare a solution of 10 mg cremophor in 50 cm^3 ethanol.

2) Place the substrate for 30 minutes in a 1M solution of KOH at 50°.

3) Place the substrate, without rinsing, directly in the boiling solution of cremophor in ethanol.

4) Rinse the substrate with distilled water and then with *iso*-propanol and dry it quickly in hot *iso*-propanol vapour.

5) Rub the substrate uni-directionally (say 20 times) with filter paper or any related material.

6) Place the glass plates once more in the boiling cremophor solution.

7) Clean the plates ultrasonically in *iso*-propanol for no more than two minutes and dry quickly in hot *iso*-propanol vapour.

CHAPTER 3

The magnetic susceptibility

As the first example of an anisotropic physical property of a nematic liquid crystal we shall discuss the magnetic susceptibility. In the SI system of units the magnetization M is defined in terms of the magnetic induction B and the magnetic field strength H by

$$M = \mu_0^{-1}B - H, \tag{3.1}$$

where μ_0 is the permeability of vacuum. Dividing this equation by $\mu_0^{-1}B$ we obtain the dimensionless quantity χ, the magnetic susceptibility:

$$\chi = \mu_0 M / B. \tag{3.2}$$

To describe the magnetization in an anisotropic phase the scalar χ must be replaced by a second-rank tensor $\boldsymbol{\chi}$, and Eq. (3.2) is generalized to

$$M_\alpha = \mu_0^{-1}\chi_{\alpha\beta}B_\beta, \qquad \alpha, \beta = x, y, z, \tag{3.3}$$

where $\chi_{\alpha\beta}$ is an element of $\boldsymbol{\chi}$. We restrict ourselves to uniaxial phases like the nematic phase, and take the director $\boldsymbol{n}$ along the z axis. In that case $\boldsymbol{\chi}$ is diagonal and takes the form

$$\begin{bmatrix} \chi_\perp & 0 & 0 \\ 0 & \chi_\perp & 0 \\ 0 & 0 & \chi_\| \end{bmatrix}$$

The subscripts $\|$ and $\perp$ are used to indicate the components parallel and perpendicular to the director, respectively. The average susceptibility is given by

$$\bar{\chi} = \tfrac{1}{3}\sum_\gamma \chi_{\gamma\gamma} = \tfrac{1}{3}(\chi_\| + 2\chi_\perp), \tag{3.4}$$

while the magnetic anisotropy is defined as

$$\Delta\chi = \chi_\| - \chi_\perp = \tfrac{3}{2}(\chi_\| - \bar{\chi}). \tag{3.5}$$

Hence the susceptibility tensor has only two different nonzero elements, and we find

$$\begin{aligned} M &= \mu_0^{-1}\chi_\| B, \quad \text{if } B \| n, \\ M &= \mu_0^{-1}\chi_\perp B, \quad \text{if } B \perp n. \end{aligned} \tag{3.6}$$

For an arbitrary angle between $\boldsymbol{B}$ and $\boldsymbol{n}$, we derive for the total magnetization

$$M = \mu_0^{-1}\chi_\perp \boldsymbol{B} + \mu_0^{-1}\Delta\chi(\boldsymbol{B}\cdot\boldsymbol{n})\boldsymbol{n}. \tag{3.7}$$

The free energy in a magnetic field (Guggenheim, 1967, Ch. 11) then is given by

$$F_{\mathrm{magn}} = -\int \boldsymbol{B}\cdot d\boldsymbol{M} = -\tfrac{1}{2}\mu_0^{-1}\chi_\perp B^2 - \tfrac{1}{2}\mu_0^{-1}\Delta\chi(\boldsymbol{B}\cdot\boldsymbol{n})^2. \tag{3.8}$$

The susceptibility has been derived from the magnetization, *i.e.* the magnetic moment per unit volume. Hence χ is the volume susceptibility, and F_{magn} in Eq. (3.8) is the magnetic free energy density. In addition to the volume susceptibility χ, one can introduce the mass susceptibility χ^m. If the sample has mass m and volume V, we have $\chi^m m = \chi V$, or using the density ρ:

$$\chi^m = \chi/\rho \quad (\mathrm{m^3 kg^{-1}}). \tag{3.9}$$

The molar susceptibility χ^M refers to a mole of substance:

$$\chi^M = \chi^m M \quad (\mathrm{m^3 mol^{-1}}), \tag{3.10}$$

where M is the mass number. Like most organic substances, liquid crystals are usually diamagnetic. Consequently $\chi_\parallel$ and $\chi_\perp$ are small and negative, of the order of 10^{-5} SI units.[†] In most cases $\Delta\chi$ is positive. As diamagnetism is independent of the temperature, $\bar{\chi}^m$ in the nematic phase is equal to the isotropic susceptibility χ^m, measured above T_{NI}. Recently some paramagnetic mesomorphic free radicals have been synthesized (Dvolaitzky *et al.* 1976).

The magnetic susceptibility is conveniently measured by means of the classical Faraday-Curie method. A sample of volume V is placed in a strong magnetic field and acquires a magnetization $\boldsymbol{M}$. In the nematic phase, the director will be aligned along the field (see Section 2.2) and the magnetization is proportional to $\chi_\parallel$. In addition to the main field a small gradient is applied along the vertical direction (chosen as the x axis, the z axis being along the director). Due to its magnetic moment, the sample experiences a force K in the direction of this gradient, given by (see, for example, Weiss and Witte, 1973)

$$K = \mu_0^{-1}(\chi-\chi_1)\int_V (\partial B^2/\partial x)\,dV = \mu_0^{-1}(\chi-\chi_1)V(\partial B^2/\partial x)_{\mathrm{av}}. \tag{3.11}$$

Here χ is the volume susceptibility of the sample, and χ_1 that of the gas

---

[†]In SI units χ is a factor of 4π larger that in CGS units; for χ^m and χ^M additional factors 10^{-3} occur (Mulay, 1976).

driven out by the sample. If this gas itself is diamagnetic, χ_1 is small and independent of temperature, and can be disregarded. Using the mass susceptibility, Eq. (3.11) can then be written as

$$K = \mu_0^{-1}\chi^m m(\partial B^2/\partial x)_{av}. \tag{3.12}$$

The force K and the mass m can be measured directly with a sensitive balance. The values obtained have to be corrected for those of the empty sample holder. The quantity $(\partial B^2/\partial x)_{av}$ can be determined by applying the same procedure to a reference liquid. If $\partial B^2/\partial x$ is not constant over the sample, the reference is made to occupy, as closely as possible, the same volume as the substance with unknown susceptibility at the same position between the poles of the magnet. Then to a very good approximation

$$\chi^m = (K/K_{ref})(m_{ref}/m)\chi_{ref}^m. \tag{3.13}$$

Often the measurements are carried out using a helium atmosphere. This is not strictly necessary, but in the case of an air atmosphere, χ_1 is relatively large and temperature dependent because air is paramagnetic (owing to its oxygen content). Then we have to go back to Eq. (3.11) to calculate corrections to Eq. (3.13). Although these corrections are still small (a few per cent), they are important in the case of nematic liquid crystals, because of the temperature dependence of χ_{air}. Experimental details have been given by, for example, Regaya and Gasparoux (1971), and de Jeu and Claassen (1978). Although with the nematic phase, necessarily, only $\chi_\parallel$ can be measured, $\bar{\chi}$ can be obtained for the isotropic phase, and with Eq. (3.5) full information on $\Delta\chi$ is available. Figure 3.1 shows the magnetic susceptibility of PAA, for which compound the results of various authors agree very well. For MBBA, there is considerable disagreement amongst the results of different authors, as discussed by de Jeu *et al.* (1976). Rotating field methods have also been used to determine $\Delta\chi$ (Gasparoux and Prost, 1971). However, the values thus obtained are in general too low.

The anisotropy of a physical property such as the magnetic susceptibility can be used as an order parameter to describe the degree of orientational order of the nematic phase. It fulfills the requirements of an order parameter, as the anisotropy vanishes above T_{NI} and acquires a finite value below the phase transition. The magnetic susceptibility is a convenient choice of property because the diamagnetic moments of the molecules are very small, and hence interactions between these moments can be ignored. This can be concluded from the fact that the relative diamagnetic permeability $\mu_r = 1 + \chi$ differs only very little from unity. Consequently, the field working on a molecule can be taken as equal to the externally applied

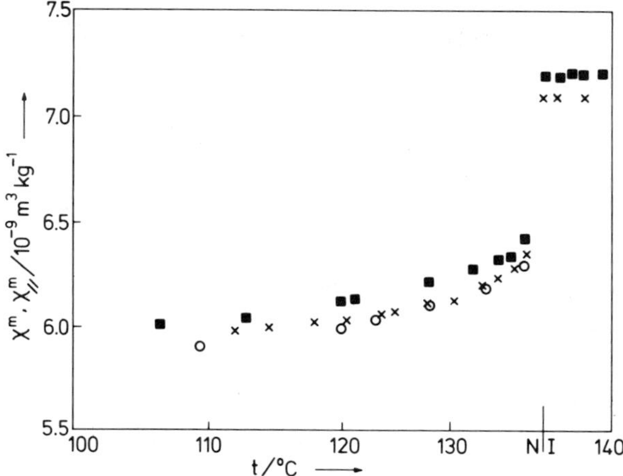

FIGURE 3.1   Magnetic susceptibility of PAA; ■ de Jeu *et al.* (1976), × Gasparoux *et al.* (1971), ○ Föex (1933).

field, and this facilitates a molecular interpretation. Often an order parameter $\mathbf{Q}$ is defined in a formal way by extracting the anisotropic part of χ (de Gennes, 1974):

$$Q_{\alpha\beta} = \chi_{\alpha\beta} - \delta_{\alpha\beta}\bar{\chi}, \qquad \alpha,\beta = x,y,z. \tag{3.14}$$

Here, $\delta_{\alpha\beta}$ is the Kronecker delta that has the value one if $\alpha = \beta$, and otherwise is zero. $\mathbf{Q}$ is a second-rank tensor, which is diagonal if we choose $\boldsymbol{n}$ along the z axis, has zero trace, and vanishes for the isotropic phase. In the case of uniaxial symmetry around $\boldsymbol{n}$ it is therefore sufficient to consider only one element of $\mathbf{Q}$, for example

$$Q_{zz} = \tfrac{2}{3}(\chi_\parallel - \chi_\perp). \tag{3.15}$$

In fact, all relevant information is contained in $\Delta\chi$.

In order to discuss the molecular interpretation of χ, and to relate $\mathbf{Q}$ to the microscopic order parameter S, let κ be the tensor of the molecular magnetic polarizability. The tensor κ is diagonal in a molecule-fixed coordinate system ξ,η,ζ, with the ζ axis as the long molecular axis. If we approximate the molecules by rigid rods with cylindrical symmetry (model 1 of Chapter 1), the diagonal elements of κ are $\kappa_l = \kappa_{\zeta\zeta}$ and $\kappa_t = \tfrac{1}{2}(\kappa_{\xi\xi} + \kappa_{\eta\eta})$. In that case a molecular interpretation of $\mathbf{Q}$ is particularly simple. If we apply a magnetic field along the director, a magnetic moment $\boldsymbol{m}$ is induced in a molecule. Denoting the angle between the molecular ζ axis and $\boldsymbol{n}$ by θ, the components of m are:

$$m_l = \mu_0^{-1} \kappa_l B \cos\theta,$$
$$m_t = \mu_0^{-1} \kappa_t B \sin\theta. \tag{3.16}$$

This, in turn, gives a magnetic moment along the z axis

$$m_{\parallel} = \mu_0^{-1} \left(\kappa_l \cos^2\theta + \kappa_t \sin^2\theta \right) B. \tag{3.17}$$

Hence the average moment along the z axis is given by

$$\langle m_{\parallel} \rangle = \mu_0^{-1} \left(\kappa_l \langle \cos^2\theta \rangle + \kappa_t \langle \sin^2\theta \rangle \right) B, \tag{3.18}$$

or, expressing the averages in the order parameter S:

$$\langle \kappa \rangle_{\parallel} = \tfrac{1}{3} \left[\kappa_l (2S+1) + \kappa_t (2-2S) \right]. \tag{3.19a}$$

Similarly we find

$$\langle \kappa \rangle_{\perp} = \tfrac{1}{3} \left[\kappa_l (1-S) + \kappa_t (2+S) \right]. \tag{3.19b}$$

These expressions are not restricted to the magnetic polarizability, but give in fact quite generally the macroscopic averages of any uniaxial molecular tensor. They will be encountered in any molecular interpretation of macroscopic physical properties. If N is the number of molecules, we derive from Eq. (3.19) the susceptibilities:

$$\chi_{\parallel} = N\langle \kappa \rangle_{\parallel} = \tfrac{1}{3} \left[\chi_l (2S+1) + \chi_t (2-2S) \right],$$
$$\chi_{\perp} = N\langle \kappa \rangle_{\perp} = \tfrac{1}{3} \left[\chi_l (1-S) + \chi_t (2+S) \right], \tag{3.20}$$

where we have taken $\chi_l = N\kappa_l$ and $\chi_t = N\kappa_t$ for the principal values of the susceptibility in the case of perfect orientational order ($S=1$). Combination leads to the desired expression for Q_{zz} or $\chi_{\parallel} - \chi_{\perp}$:

$$\chi_{\parallel} - \chi_{\perp} = (\chi_l - \chi_t) S, \tag{3.21}$$

which is Tsvetkov's expression for the order parameter (Saupe and Maier, 1961). Note that $\chi_{\parallel} - \chi_{\perp}$ and $\chi_l - \chi_t$ refer to the same temperature and density. This is automatically taken into account if we use the analogous expression in terms of the molar susceptibilities.

When the molecules cannot be considered to be effectively axially symmetric, the situation is slightly more complicated. Taking model 2 of Chapter 1, in which the molecules still possess two perpendicular symmetry planes, we need, besides θ, the second Euler angle ψ to specify the orientation of a molecule. Eq. (3.17) now has to be replaced by

$$m_{\parallel} = \mu_0^{-1}\left[\kappa_{\zeta\zeta}\cos^2\theta + \left(\kappa_{\xi\xi}\sin^2\psi + \kappa_{\eta\eta}\cos^2\psi\right)\sin^2\theta\right]B$$
$$= \mu_0^{-1}B\sum_i \kappa_{ii}i_z^2, \qquad i = \xi,\eta,\zeta, \tag{3.22}$$

where i_z is the z component of a unit vector along the molecular ξ, η or ζ axis, respectively. This leads to

$$\chi_{\parallel} = N\sum_i \kappa_{ii}\langle i_z^2\rangle. \tag{3.23}$$

The magnetic anisotropy or Q_{zz} is then given by

$$Q_{zz} = N\sum_i \kappa_{ii}\langle i_z^2 - \tfrac{1}{3}\rangle$$

$$= \tfrac{2}{3}N\sum_i \kappa_{ii}S_{ii}, \tag{3.24}$$

where the generalized order parameter S_{ii} has been introduced:

$$S_{ii} = \langle \tfrac{3}{2}i_z^2 - \tfrac{1}{2}\rangle, \qquad i = \xi,\eta,\zeta. \tag{3.25}$$

As **S** has zero trace, there are two independent scalar order parameters, for which, in agreement with Chapter 1, we choose $S = S_{\zeta\zeta}$ and $D = S_{\eta\eta} - S_{\xi\xi}$. Using Eq. (3.15) for Q_{zz}, we then obtain from Eq. (3.24) (Alben *et al.* 1972):

$$(\chi_{\parallel} - \chi_{\perp})/N = \left[\kappa_{\zeta\zeta} - \tfrac{1}{2}(\kappa_{\xi\xi} + \kappa_{\eta\eta})\right]S + \tfrac{1}{2}(\kappa_{\eta\eta} - \kappa_{\xi\xi})D. \tag{3.26}$$

This result reduces to Eq. (3.21) if either $\kappa_{\xi\xi} = \kappa_{\eta\eta}$ or $D = 0$.

A measurement of $\chi_{\parallel} - \chi_{\perp}$ is sufficient to determine the order parameter Q_{zz} that distinguishes the nematic phase from the isotropic phase. However, from Eq. (3.26) we note that to characterize the average degree of orientation of the molecules, we need in principle two parameters, which cannot both be determined from a single experiment. Only if we assume $D = 0$ is the anisotropy of the diamagnetic susceptibility directly proportional to the microscopic order parameter S. In order to determine whether D is indeed very small we have to combine magnetic measurements with other methods such as, for example, nuclear magnetic resonance (NMR). Let us consider PAA and the next higher homologue, *p*-azoxyphenetole (PAP). From the *ortho*-proton-proton coupling S has been determined as a function of temperature (Rowell *et al.* 1965). This coupling depends on the inter proton-proton vector. Because the angle between this vector and the long molecular axis is small, the result for S is practically independent of D. From Figure 3.2 we see that there is good agreement with S calculated

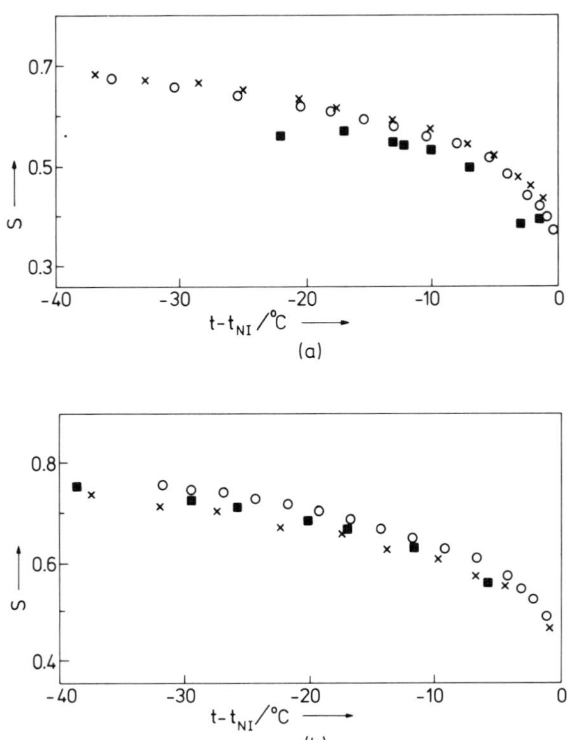

FIGURE 3.2 Order parameters for PAA (a) and PAP (b); ✕ magnetic susceptibility, ■ proton-proton coupling, (Rowell *et al.* 1965), ○ *para* carbon-13 chemical shift. (Pines and Höhener, 1976).

from the magnetic susceptibilities using Eq. (3.21) and a value of $\chi_l - \chi_t = 0.69 \times 10^{-9}$ m³ mol⁻¹. This indicates that within the experimental accuracy D can be taken as very small. In addition we consider the *para* carbon-13 chemical shift (Pines and Chang, 1974), which depends not only on S but also strongly on D, due to the anisotropy of the shielding tensor of benzene. Assuming $D = 0$, again good agreement is obtained with S as determined from the other methods (see Figure 3.2). These results depend on the relative accuracy of two independent measurements, which is necessarily not very good. More reliable results can be obtained from NMR measurements, where in a single experiment various coupling constants can be obtained. Preliminary experiments in this direction indicate values of D that are indeed zero or at most very small (Emsley *et al.* 1978; Höhener *et al.* 1979). Therefore we shall disregard D in the remaining part of this book.

When calculating S of PAA and PAP from the diamagnetic anisotropy we assumed a value of $\chi_l - \chi_t = 0.69 \times 10^{-9}$ m³ mol⁻¹, which was chosen to ensure good agreement with the NMR data for both compounds. In principle $\chi_l - \chi_t$ can be determined from measurements made on a solid single crystal. The only nematogenic compound for which such measurements have been performed is PAA (Föex, 1933), which crystallizes in the monoclinic system (Krigbaum et al. 1970; Carlisle and Smith, 1971). The ζ axis will be identified with the axis through the outer carbon atoms of the azoxybenzene part of the molecule. For different molecules this axis makes angles of $\pm 6.1°$ with the mirror plane of the unit cell, while its projections on this plane are parallel to each other (see Figure 3.3). Then the direction of the largest susceptibility is in the mirror plane at an angle of $8.8°$ to the ζ axis. If the PAA crystal is approximated by a uniaxial system we obtain the value of $\chi_l - \chi_t = 0.77 \times 10^{-9}$ $m^3\,mol^{-1}$. As this value is obtained as the difference of two much larger quantities, the accuracy is estimated to be at best 10%. In Table 3.1 this result is compared with that of biphenyl and with that of some other substances having two benzene rings that are related to classes of well-known nematogenic compounds. In most cases we find a value of $\chi_l - \chi_t$ somewhat below twice the value for benzene, PAA being the only exception. The value for $\chi_l - \chi_t$ of PAA and PAP used in Figure 3.2, although somewhat smaller than the solid state value of PAA, compares favourably with the general trend.

Finally we note that another method has been in use to estimate $\chi_l - \chi_t$. A logarithmic plot of $\chi_\parallel - \chi_\perp$ against temperature often gives a straight line, and $\chi_l - \chi_t$ is obtained by extrapolating to $T = 0$ and assuming $S = 1$ at this point (Haller et al. 1973). This assumption implies that a formula of the form of Eq. (1.6), which has no theoretical justification, is valid down to $T = 0$. We found that for the series of PAA and higher homologues $\chi_l - \chi_t$ thus obtained fluctuates with increasing chain length. As the contribution of the paraffinic chains to the susceptibility anisotropy is rather small, this indicates that the results obtained cannot be expected to be accurate.

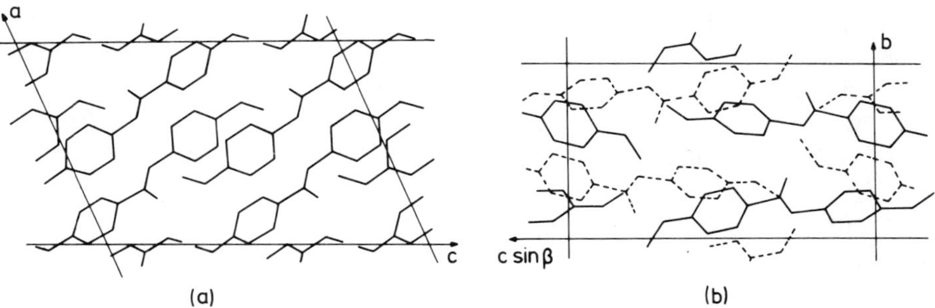

FIGURE 3.3 Crystal structure of PAA; projection on the plane normal to the two-fold axis (a) and looking down the c-axis (b) (Carlisle and Smith, 1971).

TABLE 3.1

Molecular diamagnetism of some compounds (10^{-9} $m^3\,mol^{-1}$) (Pacault *et al.* 1967)

Compound	$\chi_{\zeta\zeta}$	$\chi_{\xi\xi}$	$\chi_{\eta\eta}$	$\chi_l - \chi_t$
(benzene ring structure)	−0.438	−0.438	−1.189	0.374
(biphenyl structure)	−0.851	−0.775	−2.310	0.690
(terphenyl structure)	−1.216	−1.107	−3.409	1.042
(diphenylacetylene structure)	−1.024	−0.852	−2.492	0.648
(diphenyldiacetylene structure)	−1.375	−0.946	−2.597	0.397
(stilbene/bibenzyl structure)	−1.141	−1.141	−2.542	0.701
(stilbene structure)	−1.078	−0.630	−2.634	0.553[a]
(azobenzene structure)	−0.985	−0.679	−2.430	0.569[a]
CH_3O—(ring)—N=N(→O)—N=—(ring)—OCH_3 (azoxyanisole structure)	−1.330	−1.130	−3.079	0.774

[a] There are some uncertainties about the molecular stereochemistry in the solid phase, which make these results less reliable (Lumbroso-Bader, 1956).

Diamagnetism is due to changes in the precession of the electrons around the nuclei under the influence of a magnetic field. This leads to an induced magnetic moment that counteracts the field from which it originates, and hence to a negative sign of χ. As atomic susceptibilities are isotropic one would expect little anisotropy of the susceptibility in the case of molecules as well. However, in aromatic systems, a strong anisotropy is observed (see, for example, Haberditzl, 1976). Qualitatively this can be attributed to the fact that the π-electrons in a benzene ring are delocalized and form a "ring-current". Consequently, when the magnetic field is perpendicular to the plane of the ring a large counteracting induced moment can be expected (compare $\chi_{\eta\eta}$ in Table 3.1). According to Eq.

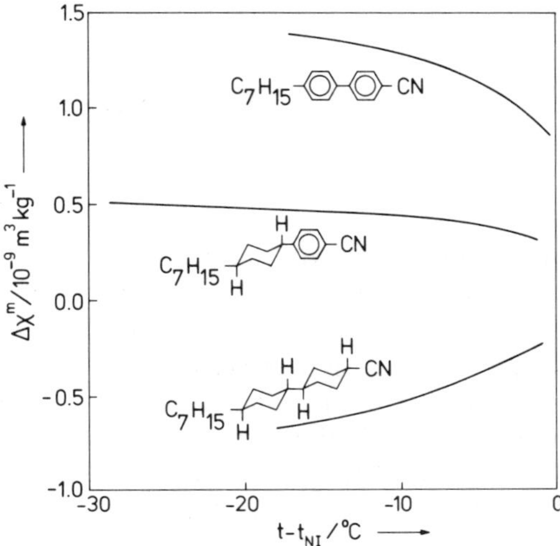

FIGURE 3.4 Magnetic anisotropy of some compounds with different numbers of benzene rings.

(3.21) we may then expect in nematic liquid crystals a positive anisotropy of the susceptibility that is proportional to the number of aromatic rings. This proportionality was in fact observed by Tsvetkov and Sosnovsky (1943). It is once more illustrated in Figure 3.4 for some more recently synthesized compounds. For a compound with two benzene rings $\Delta\chi^m$ is of the order of 10^{-9} m^{3}kg^{-1}. For each benzene ring that is replaced by a cyclohexane ring the anisotropy of the susceptibility drops. When no aromatic ring is left, $\Delta\chi^m$ has become negative. This is due to the negative anisotropy of the C$\equiv$N bond (Flygare, 1974). From Table 3.1 we see that an analogous effect is observed for other triple bonds. Especially the $-C\equiv C-C\equiv C-$ bridge decreases $\chi_l - \chi_t$ appreciably. Qualitatively this effect can be attributed to ring currents perpendicular to the long axis, which are possible in the case of triple bonds. Consequently combination of this bridge with cyclohexane rings and CN groups could lead to nematic liquid crystals with a relatively large negative diamagnetic anisotropy.

CHAPTER 4

The refractive index

A uniaxial (liquid) crystal has two principal refractive indices, n_o and n_e. The first one, n_o, is observed for an "ordinary" ray, associated with a light wave where the electric vector vibrates perpendicular to the optical axis. The "extraordinary" index, n_e, is observed for a linearly polarized light wave where the electric vector is parallel to the optical axis. In the case of a nematic or a uniaxial smectic liquid crystal the optical axis is given by the director. Still using the subscripts $\parallel$ and $\perp$ for the directions parallel and perpendicular to the director, we then have

$$n_o = n_\perp,$$
$$n_e = n_\parallel. \tag{4.1}$$

The birefringence is given by

$$\Delta n = n_e - n_o = n_\parallel - n_\perp. \tag{4.2}$$

In practice, we find $n_\parallel > n_\perp$; Δn is therefore positive (see Figure 4.1a) and varies from values close to zero to about 0.4. Note that Eq. (4.1) is not valid for a chiral nematic liquid crystal. In that case, the uniaxial optical axis is the helix axis, which is perpendicular to the local director. When the wavelength of the light is much larger than the pitch, we find for chiral nematics:

$$n_o = \left[\tfrac{1}{2}\left(n_\parallel^2 + n_\perp^2 \right) \right]^{\frac{1}{2}},$$
$$n_e = n_\perp. \tag{4.3}$$

As one still usually has $n_\parallel > n_\perp$ this leads to $\Delta n = n_e - n_o < 0$; chiral nematics then have a negative birefringence (see Figure 4.1b).

More formally, we note that the refractive index is related to the response of matter to an electric field. On application of a field E, an electric polarization P is induced that is related to E and to the dielectric displacement D by

$$P = D - \varepsilon_0 E, \tag{4.4}$$

where ε_0 is the permittivity of vacuum. For small fields, P is proportional to E. In the case of an anisotropic medium like a uniaxial (liquid) crystal

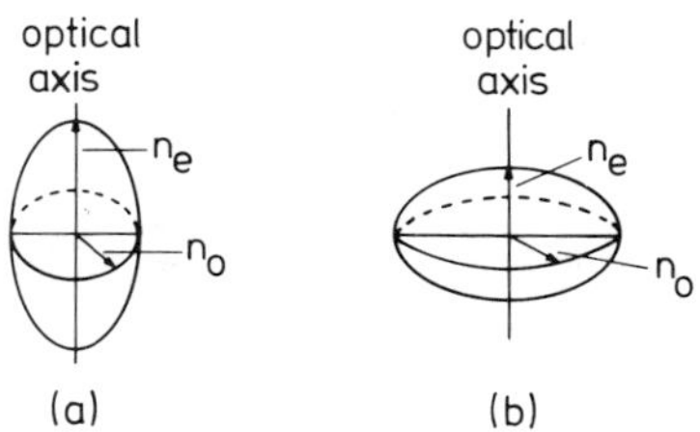

FIGURE 4.1   Polarizability ellipsoid of a positive uniaxial (a) and a negative uniaxial (b) medium.

we have Eq. (4.5) which is similar to Eq. (3.3) of the magnetic case,

$$P_\alpha = \varepsilon_0 \chi^e_{\alpha\beta} E_\beta, \qquad \alpha, \beta = x, y, z, \tag{4.5}$$

where $\chi^e_{\alpha\beta}$ is now an element of the electric susceptibility tensor $\boldsymbol{\chi}^e$. In the electric case it is usually preferable to work with the permittivity ε, which is related to the susceptibility by

$$\varepsilon = \mathbf{I} + \boldsymbol{\chi}^e, \tag{4.6}$$

where $\mathbf{I}$ is the unit tensor. In general, ε will depend on the frequency of the applied field. In this chapter, we are interested in the optical frequency range, where

$$\varepsilon_\alpha = n^2_\alpha, \qquad \alpha = x, y, z, \tag{4.7}$$

n_α being the refractive index along the α direction. Equations (4.5–4.7) give the phenomenological relation between the refractive index and the electric polarization. Again taking the director along the z axis, we have $n_\parallel = n_z$ and $n_\perp = [\frac{1}{2}(n^2_x + n^2_y)]^{\frac{1}{2}}$.

Measurements of the refractive indices of a liquid crystal are conveniently carried out by Abbe's double prism method. In this method, the liquid crystal is used as a thin film between the hypotenuse areas of two prisms. If the refractive index of the prism is greater than the indices of the liquid crystal, the boundary angles corresponding to total reflection of the ordinary as well as the extraordinary ray can be measured (see Figure 4.2). The boundary associated with n_o can be made more distinct by blocking the extraordinary light with a polarizer. In practice it suffices to use a commercial refractometer, which provides a direct reading of n_o. Usually $n_o < n_{\text{prism}} < n_e$, and n_e cannot be determined unless special prisms are used. Alternatively, instead of measuring n_e, one can determine Δn directly using interference techniques.

In order to determine Δn, one can use a uniform planar sample with its optical axis at an angle of 45° to the crossed polarizers. Normally incident

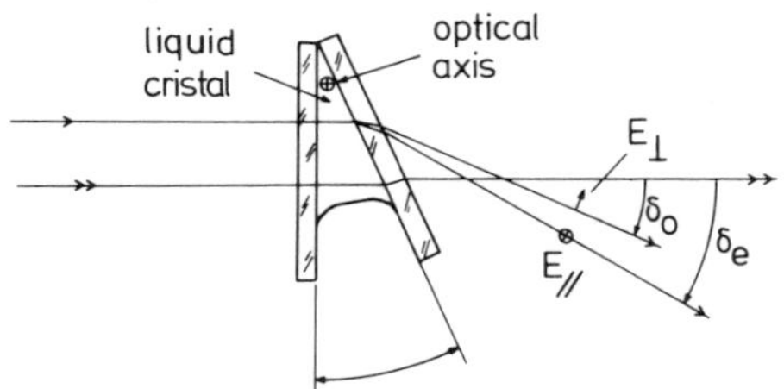

FIGURE 4.2 Angle of total reflection of the ordinary and the extraordinary ray in a wedge-shaped sample.

light is split into an ordinary and an extraordinary ray of equal intensity. After passage through the sample, the two rays have a phase difference and are made to interfere at the analyzer. The intensity of the transmitted light is monitored with a photodetector. If λ is the wavelength of the (monochromatic) light, and d the thickness of the sample, the intensity is a minimum when

$$d\Delta n = k\lambda, \qquad k \text{ is an integer.} \tag{4.8}$$

In order to determine Δn, Eq. (4.8) can be used in various ways.

a) With a fixed d and λ, the intensity of the transmitted light can be recorded as a function of the temperature, which is varied continuously. This provides a direct measurement of $\Delta n(T)$. However, at the NI transition, Δn varies discontinuously. Consequently, the value of k at the first minimum below $T_{\rm NI}$ is not known, and one additional independent measurement is necessary to fix the absolute scale of Δn.

b) At fixed values of λ and T, a wedge-like sample can be used in which d varies along the x axis from 0 to a value d_0 (Haller *et al.* 1972). If d_0 is reached at a distance x_0 from the apex, the thickness at a point x along the wedge is given by $d = xd_0/x_0$. In this situation a series of uniformly spaced fringes appear, whose separation Δx can be measured. As successive fringes correspond to a change in k of ± 1, we find

$$\Delta n = \lambda x_0/(\Delta x d_0). \tag{4.9}$$

This method has the advantage that Δn is determined absolutely. However, the measurement is more elaborate than that in method (a) and must be repeated at each desired temperature.

c) For a fixed thickness and temperature, the wavelength can be varied. If, at a certain value of k, a minimum in the transmitted light occurs at a wavelength λ_1, on decreasing the wavelength, the next minimum is found at $(k+1)\lambda_2$. Hence, in this way, the relative variation of Δn with wavelength can be determined. In order to fix the absolute scale, the value of k

must be known, and again one independent measurement of Δn is necessary.

In general it seems that method (a) at several fixed wavelengths is the simplest way to obtain full information on Δn. The absolute scale can be fixed as follows. Disregarding for a moment any discontinuity at T_{NI} one can extrapolate n_{is} to lower temperatures to find an approximate value for $\bar{n} = \frac{1}{3}(n_{\parallel} + 2n_{\perp})$. From $\bar{n}$ and $n_{\perp}$ an approximate value of Δn can be estimated. For a thin sample with k of the order of 5 the accuracy of this procedure suffices to determine k unambiguously (Leenhouts and van der Woude, 1978). Alternatively, method (b) can be used for one fixed temperature.

Finally, we mention that interference techniques can be used in a different way. When the back substrate is made reflecting, and the surface at the side of the liquid crystal is made half reflecting, the interference of the light reflected at the different boundaries can be observed. For a planar sample, depending on the polarization direction of the incident beam, the variation of $n_{\parallel}$ or $n_{\perp}$ with for example temperature or pressure, is thus obtained. In general it will be preferable to measure Δn directly, but this method has proved useful for high-pressure studies of refractive indices (Horn, 1978b).

In Figure 4.3 we give some results for the refractive indices of PAA at various wavelengths. In the nematic phase both refractive indices show normal dispersion (n decreases with increasing λ), which is, however, much larger for $n_{\parallel}$ than for $n_{\perp}$. Figure 4.4 gives some results for Δn for the series of p,p'-dialkoxyazoxybenzenes. Within the homologous series, there is a pronounced alternation of Δn, which is parallel to the well-known alterna-

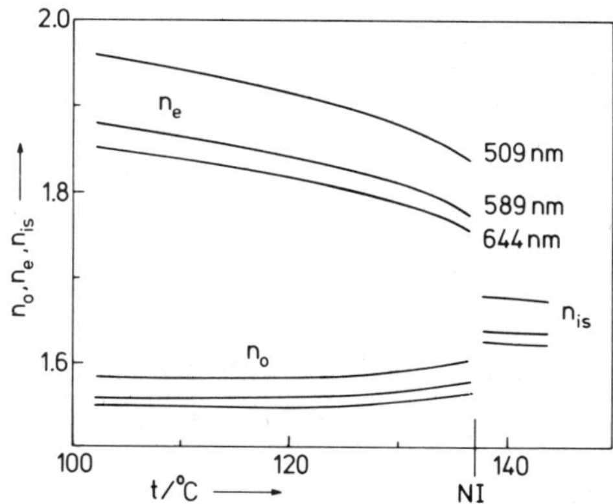

FIGURE 4.3   Refractive indices of PAA at various wavelengths (Chatelain and Germain, 1964).

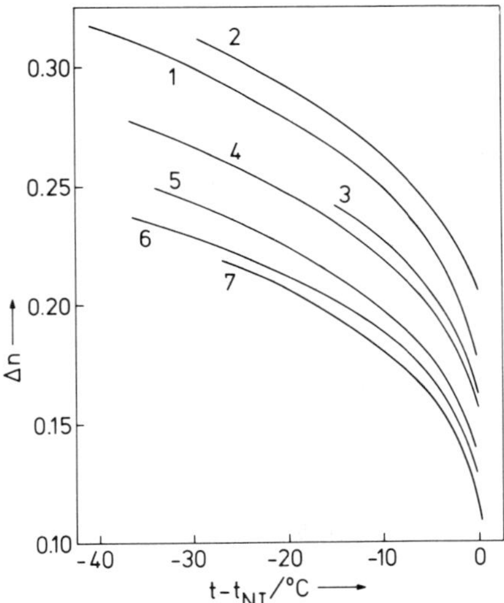

FIGURE 4.4   Birefringences of members of the series of p,p'-dialkoxyazoxybenzenes at 633 nm (Hanson and Shen, 1976); the number of carbon atoms in the alkoxy chain is indicated.

tion of T_{NI}. Measurements of refractive indices of liquid crystals are relatively numerous. Results can be found in the literature for nematic azoxybenzenes (Chatelain and Germain, 1964; Hanson and Shen, 1976; de Jeu and Bordewijk, 1978), alkoxybenzoic acids (Ryumtsev *et al.* 1974), tolanes (Labrunie and Bresse, 1973), Schiff's bases (Brunet-Germain, 1970; Pelzl *et al.* 1975), cyanobiphenyls (Horn, 1978a), phenyl benzoates (Pelzl *et al.* 1975; Schulze and Burkersrode, 1975), and cholesteryl esters (Pelzl and Sackmann, 1973). See also Pelzl and Sackmann (1971), and Haller *et al.* (1973).

When a body is placed in a uniform electric field E of moderate intensity, a dipole moment m is induced, which has components that depend linearly on E (Böttcher, 1973):

$$m = \alpha \cdot E, \tag{4.10}$$

where α is the polarizability tensor of the body. An important model example is a spheroidal body of permittivity ε in vacuo, with principal axis $2a$ and $2b$. Taking a prolate spheroid with its long axis $2a$ in the direction of the z axis, the tensor α is diagonal, with elements[†]

$$\alpha_\gamma = \varepsilon_0(\varepsilon - 1)\left[1 + (\varepsilon - 1)\Omega_\gamma\right]^{-1} \tfrac{4}{3}\pi a b^2, \qquad \gamma = a,b. \tag{4.11}$$

The shape factors or depolarizing factors Ω_γ depend only on the axial

---

[†]In SI units α is expressed in Fm^2; the reduced polarizability α/ε_0 has the dimension of a volume (m^3) and equals $4\pi \times 10^{-6}$ times the polarizability in CGS units (cm^3).

ratio a/b and are given by

$$\Omega_a = 1 - w^2 + \tfrac{1}{2}w(w^2 - 1)\ln\left[(w+1)/(w-1)\right],$$
$$\Omega_b = \tfrac{1}{2}(1 - \Omega_a), \tag{4.12}$$

while $w^2 = a^2/(a^2 - b^2)$. In the case where the spheroid reduces to a sphere $(a = b)$ we have $\Omega_a = \Omega_b = \tfrac{1}{3}$, and Eq. (4.11) reduces to the equation for the scalar polarizability of a dielectric sphere of radius a and permittivity ε in vacuo.

In general we must attribute to a molecule a polarizability tensor $\boldsymbol{\alpha}$. The polarization thus induced consists of two parts. The electric field will induce a displacement of the electrons relative to the nucleus in each atom (electronic polarization) and also a displacement of the nuclei relative to one another (ionic polarization). For organic liquids the second contribution is usually only 5–10% of the first. Moreover the ionic polarization plays a role only up to frequencies in the infrared region, and is usually no longer important in the visible part of the spectrum. We shall therefore relate the refractive index, or rather the optical dielectric tensor ε, to the molecular electronic polarizability $\boldsymbol{\alpha}$. In the isotropic phase, we find, using Eq. (4.10)

$$\boldsymbol{P} = N\alpha\boldsymbol{E}_i, \tag{4.13}$$

where N is the number of molecules per unit volume and $\boldsymbol{E}_i$ is the internal field, the average field that acts on a molecule. $\boldsymbol{E}_i$ is equal to the macroscopic field $\boldsymbol{E}$ plus the average field due to the induced moments of the surrounding particles. The calculation of $\boldsymbol{E}_i$ is one of the main problems associated with the theory of electric polarization. In practice, two different approaches have been used to calculate $\boldsymbol{E}_i$ (Böttcher, 1973). The first method starts from the electric field acting on a single molecule. For the calculation of this field the environment of the molecule is considered to be a continuum with the macroscopic properties of the dielectric under consideration. The assumption of a continuum environment for a molecule is, of course, rather crude, and leads to fundamental problems concerning the applicability of results derived for macroscopic bodies to particles of molecular size. Alternatively, one can use statistical mechanics to calculate $\boldsymbol{E}_i$, taking the interactions between the molecules explicitly into account. Although this approach is *a priori* more satisfactory, in practice, the calculations can only be performed if considerable simplifications are made. We shall consider mainly the first approach.

In the case of a liquid crystal, Eq. (4.13) can be generalized to

$$\boldsymbol{P} = N\langle \boldsymbol{\alpha}\cdot\boldsymbol{E}_i \rangle, \tag{4.14}$$

where the brackets denote the average over the orientations of all molecules. The internal field being a linear function of the macroscopic field, its relation to the macroscopic field can be represented by

$$E_i = \mathbf{K} \cdot E, \qquad (4.15)$$

where $\mathbf{K}$ is an ordinary second-rank tensor. The introduction of this tensor makes it possible to write Eq. (4.14) as

$$P = N \langle \boldsymbol{\alpha} \cdot \mathbf{K} \rangle \cdot E. \qquad (4.16)$$

On the other hand, Eqs. (4.5) and (4.6) relate P to the macroscopic permittivity. Thus we find[†]

$$\varepsilon = \mathbf{I} + (N/\varepsilon_0) \langle \boldsymbol{\alpha} \cdot \mathbf{K} \rangle. \qquad (4.17)$$

Consequently we find for the dielectric anisotropy

$$\Delta\varepsilon = n_\parallel^2 - n_\perp^2 = (N/\varepsilon_0)(\langle \boldsymbol{\alpha} \cdot \mathbf{K} \rangle_\parallel - \langle \boldsymbol{\alpha} \cdot \mathbf{K} \rangle_\perp). \qquad (4.18)$$

In order to proceed with Eqs. (4.17) and (4.18) we must know the internal field tensor $\mathbf{K}$. In general, one expects $\mathbf{K}$ to depend both on the properties and the orientation of the molecule at which the internal field is considered, and on the dielectric tensor of the macroscopic sample. Retaining all these dependencies, very complicated expressions are obtained (Kuznetsov *et al.* 1975). Instead we shall rely on some experimental information to simplify the internal field problem.

Experimental results for the permittivity at optical frequencies reveal two important facts (de Jeu and Bordewijk, 1978). Firstly, if we denote the average permittivity by $\bar{\varepsilon} = \frac{1}{3}(\varepsilon_\parallel + 2\varepsilon_\perp)$ we find that $\bar{\varepsilon}/\rho$ is a constant through the nematic and isotropic phases. Secondly, the dielectric anisotropy is directly proportional to the anisotropy of the magnetic susceptibility. As an example the latter relation is illustrated in Figure 4.5 for MBBA at various wavelengths. For axially symmetric molecules, $\Delta\chi^m$ is proportional to the order parameter S (see Eq. 3.21), and using $\Delta\chi = \rho\Delta\chi^m$ we thus find

$$\Delta\varepsilon = n_\parallel^2 - n_\perp^2 \sim \rho S. \qquad (4.19)$$

In Eq. (4.18) the particle density is given by $N = N_A \rho / M$, M being the mass number. Hence, comparison of Eqs. (4.18) and (4.19) leads to the condition for the internal field tensor

$$\langle \boldsymbol{\alpha} \cdot \mathbf{K} \rangle_\parallel - \langle \boldsymbol{\alpha} \cdot \mathbf{K} \rangle_\perp \sim S. \qquad (4.20)$$

[†]The corresponding equations in CGS units are obtained by replacing ε_0 by $1/4\pi$.

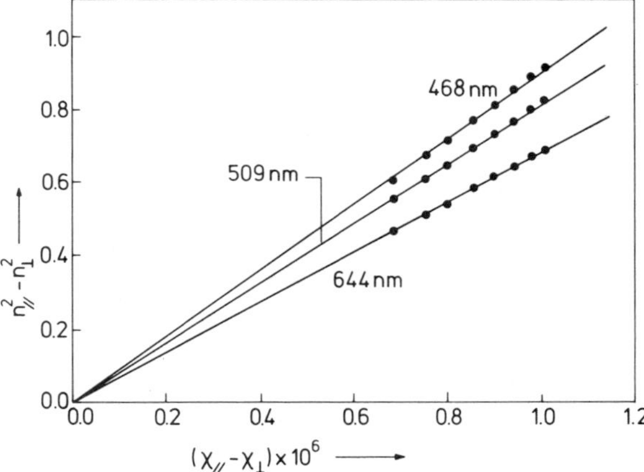

FIGURE 4.5   Proportionality between $n_\parallel^2 - n_\perp^2$ and $\Delta\chi$ for MBBA. (Optical data from Brunet-Germain, 1970; susceptibilities from de Jeu *et al.* 1976).

If we assume that the molecules are axially symmetric around the ζ axis the electronic polarizability tensor $\boldsymbol{\alpha}$ will be diagonal in the same molecular frame ξ, η, ζ that diagonalizes the magnetic polarizability tensor. Then the principal elements will be $\alpha_l = \alpha_{\zeta\zeta}$ and $\alpha_t = \frac{1}{2}(\alpha_{\xi\xi} + \alpha_{\eta\eta})$ and we find analogously to Eqs. (3.19) and (3.21)

$$\langle \boldsymbol{\alpha} \rangle_\parallel - \langle \boldsymbol{\alpha} \rangle_\perp = (\alpha_l - \alpha_t)S. \tag{4.21}$$

In order to fulfill both Eq. (4.20) and Eq. (4.21), $\mathbf{K}$ must be a molecular tensor, independent of the dielectric anisotropy of the macroscopic sample, and thus independent of the orientation of the molecule with respect to the director. It then follows that the principal axes of $\boldsymbol{\alpha}$ and $\mathbf{K}$ should coincide. As with Eq. (3.19), we then find with Eq. (4.17):

$$\begin{aligned}
\varepsilon_\parallel &= n_\parallel^2 = 1 + (N/3\varepsilon_0)\left[\,\alpha_l K_l(2S+1) + \alpha_t K_t(2-2S)\,\right], \\
\varepsilon_\perp &= n_\perp^2 = 1 + (N/3\varepsilon_0)\left[\,\alpha_l K_l(1-S) + \alpha_t K_t(2+S)\,\right],
\end{aligned} \tag{4.22}$$

where K_l and K_t are the principal values of $\mathbf{K}$. We conclude that it is possible to take the effect of the internal field into account by working with an effective or "dressed" polarizability tensor $\boldsymbol{\alpha}^* = \boldsymbol{\alpha} \cdot \mathbf{K}$. This is an important conclusion for some practical situations, such as Raman scattering (Jen *et al.* 1977). The nearest neighbours of a molecule contribute almost all of the internal field corrections because the dipole-dipole interaction falls off as r^{-3}. These near-neighbour correlations are not significantly temperature-dependent in most condensed phases, and probably do not even change over the various phase transitions. This is probably the physical origin of the observation that **K** can be taken to be a molecular tensor.

The proportionality of Eq. (4.19) has one further consequence. In combination with the constancy of $\bar{\varepsilon}/\rho$ we can derive that approximately

$$\Delta n \sim \rho^{\frac{1}{2}} S. \tag{4.23}$$

As the variation of ρ over the temperature range of a nematic is usually very small, this means that a relatively easy measurement of the temperature dependence of Δn gives a good indication of the variation of S with temperature. However, it should be borne in mind that this is not necessarily true to within the accuracy with which Δn can be obtained, but only to within the accuracy with which the proportionality $\Delta\varepsilon \sim \Delta\chi$ is valid.

If we want to calculate the molecular electronic polarizabilities α_l and α_t, we must rely on models for **K**. To obtain reasonable values in Eq. (4.22) for K_l and K_t, we consider the case of ideal ordering, $S = 1$. For that case, the values of $\varepsilon_\parallel$ and $\varepsilon_\perp$ will be approximately equal to those for the solid, if all molecules in the solid have the same orientation. Vuks (1966) suggested that the internal field in a solid with less than cubic symmetry is independent of the orientation, and is given by:

$$E_i = \tfrac{1}{3}(\bar{\varepsilon}+2)E. \tag{4.24}$$

This equation makes the tensor **K** isotropic. Substitution in Eq. (4.22) gives (Chandrasekhar and Madhusudana, 1969):

$$(\varepsilon_\parallel - 1)/(\bar{\varepsilon}+2) = (N/9\varepsilon_0)\left[\alpha_l(2S+1)+\alpha_t(2-2S)\right],$$
$$(\varepsilon_\perp - 1)/(\bar{\varepsilon}+2) = (N/9\varepsilon_0)\left[\alpha_l(1-S)+\alpha_t(2+S)\right]. \tag{4.25}$$

These equations can be used to calculate the average polarizability $\bar{\alpha} = \tfrac{1}{3}(\alpha_l + 2\alpha_t)$, giving

$$\bar{\alpha} = (3\varepsilon_0/N)(\bar{\varepsilon}-1)/(\bar{\varepsilon}+2). \tag{4.26}$$

This is the Lorenz-Lorentz equation, which is usually derived for the isotropic phase or systems with cubic symmetry, where $\bar{\varepsilon} = \varepsilon$. In that case the internal field of Eq. (4.24) is called the Lorentz field. Hence the use of Vuks' approximation leads to a formula compatible with the Lorenz-Lorentz equation for the isotropic phase.

There are a number of objections to Eq. (4.25), however. In the first place this equation lacks a molecular basis even in the case of solids, which is much simpler than the case of liquid crystals. Furthermore, in a number of cases Vuks' treatment yields rather implausible results. For PAA, both $n_\parallel$ and $n_\perp$ show normal dispersion, which is much larger for $n_\parallel$ than for $n_\perp$

(see Figure 4.3). Consequently, α_t shows anomalous dispersion when calculated using Eq. (4.25); this is clearly an artifact of the formula. Furthermore, in a iodoform crystal, the refractive index along the three-fold axis of the molecules is larger than the refractive index perpendicular to that axis, yielding, with Vuks' equation, a positive anisotropy of the polarizability, whereas measurements of the Kerr effect in solution show that the anisotropy is negative.

On the level of the Lorenz-Lorentz equation, a molecule is represented by a point polarizability in a spherical cavity. A better representation of a real molecule would be an anisotropic homogeneously polarizable spheroid filling up the cavity in an anisotropic polarized continuum. For a uniaxial solid crystal this gives (Böttcher and Bordewijk, 1978, Sec. 95)

$$\alpha_\gamma = (\varepsilon_0/N)(\varepsilon_\gamma - 1)/[1 + (\varepsilon_\gamma - 1)\Omega_\gamma], \qquad \gamma = l, t, \qquad (4.27)$$

where the Ω_γ are shape factors that depend only on the axial ratio of the spheroid. They are equal to the depolarizing factors occurring in the solution of the electrostatic problem of a dielectric spheroid in an external field—see Eq. (4.12). The corresponding internal field can be written as

$$(E_i)_\gamma = [1 + (\varepsilon_\gamma - 1)\Omega_\gamma]E_\gamma$$

$$= \varepsilon_0(\varepsilon_0 - N\alpha_\gamma\Omega_\gamma)^{-1}E_\gamma. \qquad (4.28)$$

Apart from a slight variation due to the density dependence, this equation fulfills the requirement that the internal field should not depend on the macroscopic dielectric anisotropy. With this expression for **K**, Eq. (4.22) yields (de Jeu and Bordewijk, 1978)

$$\varepsilon_\parallel = 1 + \frac{1}{3}N\left[\frac{\alpha_l(2S+1)}{\varepsilon_0 - N\alpha_l\Omega_l} + \frac{\alpha_t(2-2S)}{\varepsilon_0 - N\alpha_t\Omega_t}\right],$$

$$\varepsilon_\perp = 1 + \frac{1}{3}N\left[\frac{\alpha_l(1-S)}{\varepsilon_0 - N\alpha_l\Omega_l} + \frac{\alpha_t(2+S)}{\varepsilon_0 - N\alpha_t\Omega_t}\right]. \qquad (4.29)$$

These expressions make it possible to infer α_l and α_t from the experimental values of $\varepsilon_\parallel$ and $\varepsilon_\perp$, if the density and the order parameter are known. The results depend on the value of the a/b ratio. In Table 4.1 we give the values for the polarizabilities of PAA and p,p'-diheptylazobenzene as obtained with the spheroidal model. The former results are calculated from the solid state refractive indices of PAA (Chatelain, 1936) using Eq. (4.27); the latter results from the refractive indices in the nematic phase using Eq. (4.29) (de Jeu and Leenhouts, 1978). The long axis of the spheroid is taken as the length of a fully extended molecule, as determined from molecular

TABLE 4.1

Electronic polarizabilities for some nematogenic compounds $(10^{-40}\,\text{Fm}^2)$

λ(nm)	p-azoxyanisole		p,p'-diheptylazobenzene	
	α_l	α_t	α_l	α_t
492	—	25.2	—	—
546	77.1	25.0	114.6	44.3
589	73.5	24.8	110.5	44.2
633	—	—	107.7	44.0
650	69.9	24.6	—	—
∞	$\sim58^a$	$\sim23.9^a$	$\sim89^a$	$\sim42.8^a$

[a]Extrapolated values.

models. The width is chosen such that the volume of the spheroid is equal to the volume available to a molecule at T_{NI}.

Eq. (4.29) remains valid within the limits $S=0$ and $S=1$. For $S=0$, the result differs somewhat from the Lorenz-Lorentz equation, but for the isotropic liquid phase of an anisotropic molecular system like CS_2, it gives better agreement with experiment (de Jeu and Bordewijk, 1978). For $S=1$, Eq. (4.29) reduces, of course, to the original solid state result, Eq. (4.27). We emphasize that Eq. (4.29) is not based on a molecular model of the optical permittivity in the nematic phase, but on the experimental observations that $\bar{\varepsilon}/\rho$ is continuous at T_{NI} and that $\Delta\varepsilon\sim\rho S$. The use of Vuks' formula for the internal field, Eq. (4.24), is equivalent to Eq. (4.28) if

$$\Omega_\gamma = \tfrac{1}{3}(\bar{\varepsilon}-1)/(\varepsilon_\gamma-1), \qquad \gamma=\|,\perp. \tag{4.30}$$

If the largest value of ε_γ corresponds to the longest axis of the molecule, for which $\Omega_\gamma < \tfrac{1}{3}$, this equation is approximately valid. If the largest value of ε_γ corresponds to a short axis of the molecule (as for example with iodoform), Vuks' formula can even lead to an incorrect sign of the anisotropy of the polarizability.

Finally we mention that the Lorentz calculation of the internal field for a cubic crystal has been generalized by Neugebauer (1954). He considers an arbitrary lattice in which the molecules are represented by anisotropic point polarizabilities with parallel principal axes. This gives an equation analogous to Eq. (4.27) with factors Ω_γ that now have a different meaning and depend on the crystal structure. This difference is due to the fact that the field of a homogeneously polarizable spheroid is not equal to the field of an anisotropic point polarizability in a spheroidal cavity (Böttcher, 1973, Sec. 20).

The representation of the polarizability of a molecule by an anisotropic point polarizability is a serious limitation, since in reality, the polarizability of a molecule will be distributed over its whole volume. This makes no difference as far as spherical or nearly spherical particles are concerned, but it becomes increasingly unsatisfactory if an anisotropic shape is involved. Moreover, when Neugebauer's formula is applied to nematics, the definition of the lattice is not clear. This problem has been circumvented in an artificial way by choosing the Ω_γ such that $\bar{\alpha}$ in the nematic or solid state equals the value obtained in the isotropic phase using the Lorenz-Lorentz equation (Saupe and Maier, 1961). As this seems to be a rather unsatisfactory procedure we shall not quote results for α_l and α_t obtained in this way.

In order to test the results for the molecular electronic polarizabilities given in Table 4.1 we need values of α_l and α_t determined by an independent method. In principle, measurements of the Kerr constant can give such results. However, the interpretation of these measurements, whether for the isotropic phase or for diluted solutions, suffers from the same internal field problem as that for the refractive indices (see, for example, Böttcher and Bordewijk, 1978, Sec. 84). Hence, the results of the various methods will be consistent for any expression for the internal field that obeys Eq. (4.20), and such a consistency does not prove that the treatment of the internal field is necessarily correct. For a conclusive test we need a method in which the problem of the internal field is avoided. This is only possible with the gaseous state, but in this state a sufficient concentration of the large molecules of compounds that form liquid crystals cannot be obtained at temperatures that are low enough to prevent decomposition.

In this situation, the best way is to test for consistency using as many methods as possible. Besides measurements of the refractive indices and the Kerr constant, the dichroic ratio R as measured with optical spectroscopy is of interest in this respect. For a liquid crystal R is defined as the absorption parallel to the director divided by the absorption in the perpendicular direction. To evaluate this ratio in terms of R_{mol}, the absorption parallel to the long molecular axis divided by the absorption perpendicular to the axis, one uses (Saupe and Maier, 1961)

$$R_{\text{mol}} = \left(n_\parallel / n_\perp \right) \left[(E_i)_\perp / (E_i)_\parallel \right]^2 R. \tag{4.31}$$

Blinov *et al.* (1975) studied the absorption of some probe molecules with a structure very similar to those of the nematic solvents used. The ratio $n_\parallel / n_\perp$ increases with decreasing wavelength. On application of Neugebauer's theory, this increase would result in some enhancement of R with decreasing wavelength. On the contrary an isotropic internal field [Vuks' formula, Eq. (4.25)] and the spheroidal model (Eq. 4.29) lead to a reduction of R with decreasing wavelength. The latter result is in agreement with the observed spectral dependence of R. Secondly measurements were performed using a globular probe molecule for which $R_{\text{mol}} = 1$. In that case, R was found to be smaller than unity and, within the experimental accuracy, equal to the ratio $n_\perp/n_\parallel$. Consequently, for this particular case of an isotropic molecule in a nematic solvent the internal field is isotropic. We conclude that any anisotropy of the internal field in the nematic phase must then be associated with the anisotropy of the investigated molecules themselves. In combination with the objections to Vuks' formula given on page 42–43, this strongly supports the spheroidal model.

Reasonable estimates of the electronic polarizabilities of a molecule can often be obtained from the addition of tabulated bond polarizability data (Le Fèvre, 1965). Unfortunately, this method does not work very well for conjugated systems, where the polarizability can be much larger than the value obtained from such a summation. For example, using the polarizabilities of benzene and of the $-N=N-$ group, one calculates for azobenzene (de Jeu and Leenhouts, 1978)

$$\alpha_l = 28, \qquad \alpha_t = 21 \qquad (10^{-40}\ \text{Fm}^2).$$

From measurements of the Kerr effect for azobenzene, it has been estimated (Armstrong and Le Fèvre, 1966) that

$$\alpha_l = 38, \qquad \alpha_t = 22 \qquad (10^{-40}\ \text{Fm}^2).$$

Apart from the uncertainties in the interpretation, it is evident that the conjugation in the longitudinal direction has led to a strong increase in the value of α_l. From the polarizabilities of a $C-C$ and a $C-H$ bond one can calculate for an alkyl chain in a planar all-trans "zig-zag" conformation increments of α_l and α_t after addition of a methylene group. These results are given in Table 4.2. From the above estimates of α_l and α_t of azobenzene and these data, we calculate for p,p'-diheptylazobenzene:

$$\alpha_l = 70, \qquad \alpha_t = 49 \qquad (10^{-40}\ \text{Fm}^2).$$

The result for α_l is somewhat smaller than the value obtained from the spheroidal model given in Table 4.1. To a large extent this could be due to the uncertainties in the interpretation of the Kerr effect. On the other hand it is also possible that the anisotropy of the molecule is slightly overestimated with the procedure chosen to estimate the axes ratio a/b.

From the data in Table 4.2, we find that addition of a methylene group leads to an increase of $\alpha_l - \alpha_t$ when starting from an even number of chain

TABLE 4.2
Increments in longitudinal and transverse polarizability for an
alkyl group in the planar "zig-zag" conformation ($10^{-40}\ \text{Fm}^2$)

$C_m \rightarrow C_{m+1}$	$\delta\alpha_l$	$\delta\alpha_t$	$\delta(\alpha_l - \alpha_t)$
odd→even	1.86	2.14	−0.28
even→odd	2.59	1.76	+0.83

atoms, whereas starting from an odd number of chain atoms one expects a small decrease. This explains qualitatively the alternation of Δn that is often observed on increasing the chain length in a homologous series (see Figure 4.4). This alternation is parallel to the well-known alternation of T_{NI}. As far as the anisotropy of the dispersion forces plays a role in determining nematic behaviour [see Eq. (1.4)], a proportionality between

TABLE 4.3

Birefringences (at $T_{NI} - T = 10°C$ and 589 nm) and clearing points for some nematogenic compounds

No.	Substance	Δn	$T_{NI}(°C)$	Reference
1	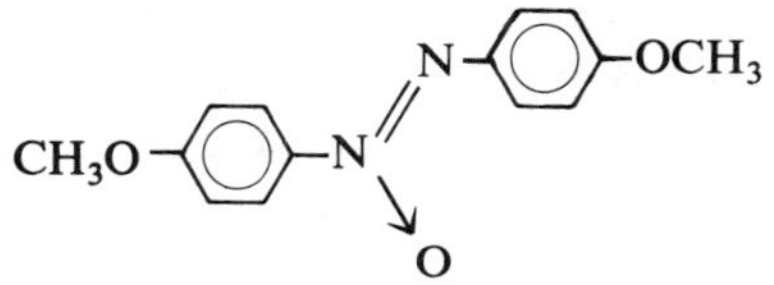 CH_3O—⟨◯⟩—N=N(→O)—⟨◯⟩—OCH_3	0.26	135	Chatelain and Germain (1964)
2	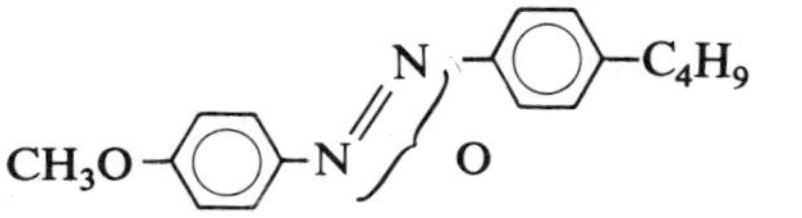 CH_3O—⟨◯⟩—N=N(→O)—⟨◯⟩—C_4H_9	0.21	76	Haller et al. (1973)
3	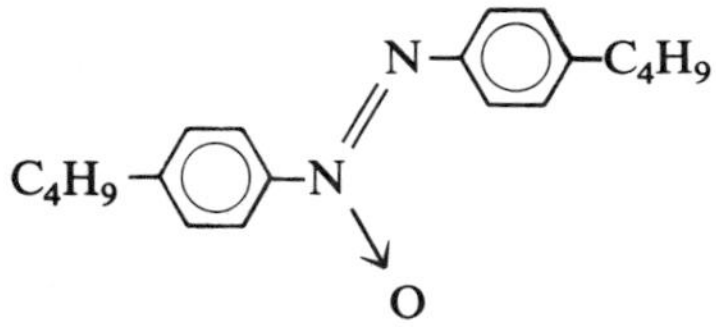 C_4H_9—⟨◯⟩—N=N(→O)—⟨◯⟩—C_4H_9	0.18	32	de Jeu and Bordewijk (1978)
4	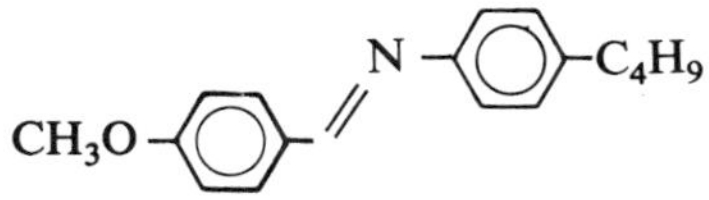 CH_3O—⟨◯⟩—CH=N—⟨◯⟩—C_4H_9	0.19	47	Brunet-Germain (1970)
5	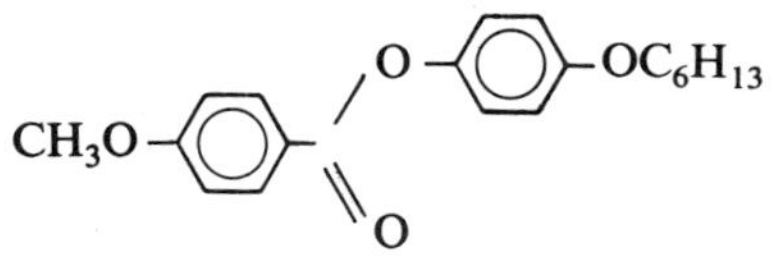 CH_3O—⟨◯⟩—C(=O)O—⟨◯⟩—OC_6H_{13}	0.13	77	Schulze and Burkersrode (1975)
6	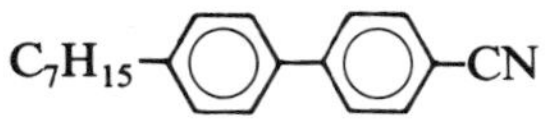 C_7H_{15}—⟨◯⟩—⟨◯⟩—CN	0.16	42	Pohl et al. (1978)
7	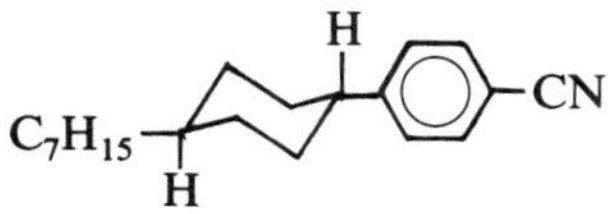 C_7H_{15}—⟨cyclohexyl⟩—⟨◯⟩—CN	0.09	57	ibid
8	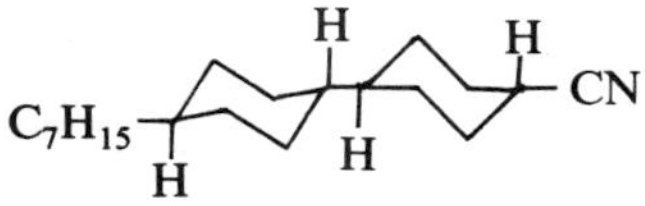 C_7H_{15}—⟨cyclohexyl⟩—⟨cyclohexyl⟩—CN	0.06	83	ibid

T_{NI} and $(\alpha_l - \alpha_t)^2$ can be expected. In general, however, other effects will certainly also come into play in determining T_{NI}. In Table 4.3 we compare some results for Δn and T_{NI} for various substances. Assuming that the density and the order parameter at $(T_{NI} - 10)°C$ are not significantly different for the various substances, Δn is a rough measure of $\alpha_l - \alpha_t$. Examples 1–3 show that when an alkyl is substituted for an alkoxy group, both Δn and T_{NI} decrease. This is to be expected, because an oxygen atom possesses lone-pair electrons that can easily become involved in the conjugated system. The decrease in α_l and thus in $\alpha_l - \alpha_t$ when an oxygen is replaced by a CH_2 group is in this case accompanied by a decrease in T_{NI}. Phenyl benzoates (no. 5) have a much less rigid bridging group than that in the other examples, and this leads to relatively small conjugative interactions. Accordingly, Δn is also smaller. In this case, however, T_{NI} is still relatively high. Finally, in examples 6–8 the benzene rings (which are rather polarizable along the plane of the ring, because of the delocalized π-electrons) are replaced by much less polarizable cyclohexane rings. As expected, the birefringence decreases strongly. Nevertheless, an increase of T_{NI} is observed. We conclude therefore that the qualitative proportionality between T_{NI} and $(\alpha_l - \alpha_t)^2$ that has been observed in some cases (de Jeu and van der Veen, 1977), does not hold in general, and is probably only valid if the changes in molecular structure that are considered are relatively minor. Then the factors other than $\alpha_l - \alpha_t$ that may influence T_{NI} (overall polarizability, anisotropic volume) cannot change appreciably.

CHAPTER 5

The dielectric permittivity

5.1 THE PERMITTIVITY IN STATIC FIELDS

Dielectric studies are concerned with the response of matter to the application of an electric field. The classical experiment is to fill a capacitor with a material and to find that the capacitance has increased from a value C to a value εC, where ε is the (relative) permittivity of the material. The increase in capacitance stems from polarization of the material by the field E, the positive charges being attracted to one electrode and the negative charges to the other end of the capacitor. For an anisotropic material the polarization (per unit volume) is given by Eq. (4.5). Using Eq. (4.6) for the permittivity, this leads to[†]

$$P = \varepsilon_0(\varepsilon - I) \cdot E. \tag{5.1}$$

In a material consisting of non-polar molecules, there is only an induced polarization, which consists of two parts: the electronic polarization (which is also present at optical frequencies) and the ionic polarization. In materials with polar molecules, there is in addition to the total induced polarization, the orientation polarization, due to the tendency of the permanent dipole moments to orientate themselves parallel to the field. In isotropic liquids and gases, the permittivity is isotropic, and in Eq. (5.1) $\varepsilon - I$ can be replaced by $\varepsilon - 1$. In solids, on the other hand, where the permittivity will usually be anisotropic, the permanent dipole moments generally have a relatively fixed orientation. In that case, the contribution to the permittivity from the orientation polarization is less important. In liquid crystals, we have the complicated situation of an anisotropic permittivity in combination with liquid-like behaviour. Hence in the case of polar molecules, there is an important contribution to the anisotropic permittivity from the orientation polarization.

We shall restrict ourselves mainly to the uniaxial liquid crystalline phases, and use a macroscopic coordinate system x, y, z, with the z axis parallel to the director. The principal elements of ε are then $\varepsilon_\parallel = \varepsilon_{zz}$ and $\varepsilon_\perp = \frac{1}{2}(\varepsilon_{xx} + \varepsilon_{yy})$. In order to measure $\varepsilon_\parallel$ and $\varepsilon_\perp$ it is common practice to use a plane capacitor, the capacitance of which is measured with AC

---

[*]The corresponding equations in CGS units are obtained by replacing ε_0 by $1/4\pi$.

excitation. The frequency (say 1 kHz) is chosen low enough to allow measurement of essentially the static permittivities and high enough to prevent electrochemical reactions or the formation of double layers at the electrodes. In practice display-like cells are often used, and consist of two glass plates, with well-defined electrodes, kept a small distance apart by spacers. The parasitic capacitance C_p is determined by means of a calibration measurement using a liquid of known permittivity as dielectric. In order to measure either $\varepsilon_\parallel$ or $\varepsilon_\perp$ one must keep the electric field parallel or perpendicular to the director as the case may be. If thin cells are made (say $\lesssim 100$ μm) this can be achieved by using either a homeotropic or a planar alignment in the cell. Transparent electrodes must then be used in order to be able to check the alignment. As a rule, however, it will be preferred to align the director with a magnetic field. This has the advantage that $\varepsilon_\parallel$ and $\varepsilon_\perp$ are measured with the same cell, thus increasing the accuracy with which $\Delta\varepsilon = \varepsilon_\parallel - \varepsilon_\perp$ is obtained. In that case non-transparent gold or copper electrodes can be used. The samples must not be too thin (say $\gtrsim 100$ μm for a magnetic induction of the order of 1 T) so that the influence of the boundary layer of the liquid crystal, where the director is not perfectly

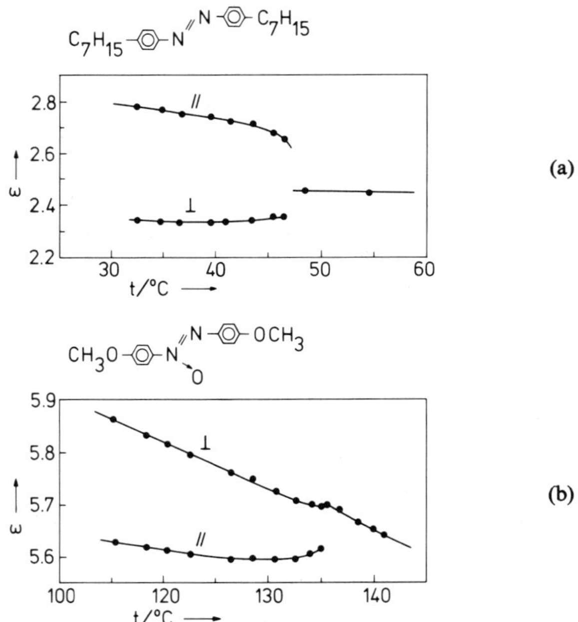

FIGURE 5.1 Dielectric permittivities of p,p'-diheptylazobenzene (a, de Jeu, 1978a) and PAA (b, Maier and Meier, 1961b).

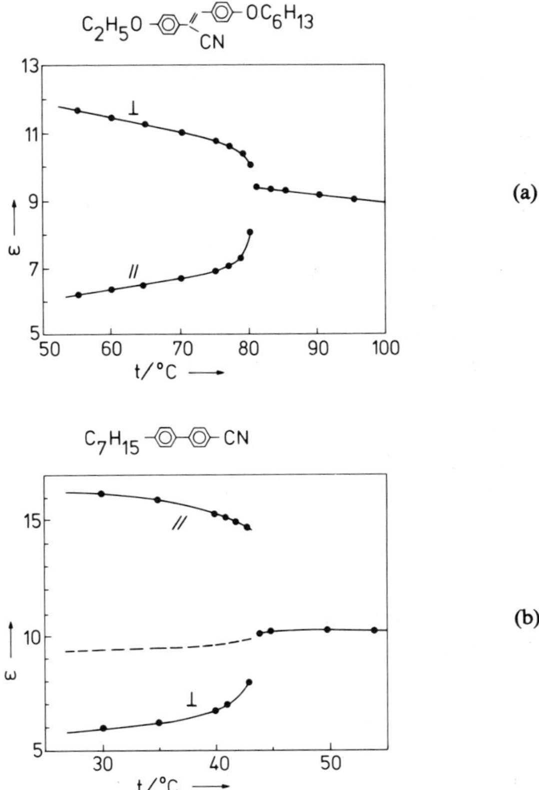

FIGURE 5.2   Dielectric permittivities of some compounds with a large negative (a, de Jeu and van der Veen, 1973) and a large positive (b, Davies *et al.* 1976; Lippens *et al.* 1977) dielectric anisotropy.

aligned by the field, can be disregarded. In all cases, the applied voltage must be small ($\lesssim 1$ V) to avoid the formation of electrohydrodynamic instabilities, which may disturb the uniform director pattern.

Figure 5.1 gives experimental results for $\varepsilon_\parallel$ and $\varepsilon_\perp$ for the nematic phase of a non-polar compound for which $\Delta\varepsilon > 0$, and for PAA, a polar compound. In the latter case the orientation polarization contributes more to $\varepsilon_\perp$ than to $\varepsilon_\parallel$, leading to a negative sign of $\Delta\varepsilon$. In practice, rather large values of $|\Delta\varepsilon|$ can be obtained, depending on the magnitude of the total dipole moment μ and the angle between μ and the long molecular axis (see Fig. 5.2). The magnitude of $\Delta\varepsilon$ usually depends on temperature. Reviews discussing the permittivity of liquid crystals have been given by Böttcher and Bordewijk (1978, Sec. 97) and by de Jeu (1978a).

A theoretical description of the permittivity of liquid crystals must show how the macroscopically observed anisotropy arises from the anisotropies of the various molecular quantities, taking account of the imperfect orientational order. Such an approach turns out to be quite feasible. It is much more difficult to proceed in the opposite way and to derive molecular quantities from the observed macroscopic permittivities. In the case of polar molecules, specific interactions between the dipole moments may occur, and this makes the situation rather complicated. Usually, moreover, there is not enough independent information on the relevant molecular properties to enable one to test conclusively the various possible approaches. Nevertheless, we shall see that dielectric studies can often provide important qualitative information on specific intermolecular interactions.

For non-polar compounds the situation is relatively simple. Using the results from Figure 5.1a, one can demonstrate that for p, p'-diheptylazobenzene $\Delta\varepsilon \sim \Delta\chi$, analogous to the situation for the optical frequency range where $n_\parallel^2 - n_\perp^2 \sim \Delta\chi$. This means that the whole reasoning of the previous chapter remains valid, and we can use Eq. (4.29) to relate ε to the molecular polarizability tensor $\boldsymbol{\alpha}$, which now, however, also includes any contribution from the ionic polarizability. In this way we find for the total static polarizabilities of p, p'-diheptylazobenzene:

$$\alpha_l = 99.3, \qquad \alpha_t = 46.8 \qquad (10^{-40}\ \mathrm{Fm^2}).$$

These values can be compared with the values obtained from the refractive index measurements given in Table 4.1, extrapolated to $\lambda = \infty$. We conclude that the contribution of the ionic polarizability to α_l and α_t is about 10 and 4 $(10^{-40}\ \mathrm{Fm^2})$, respectively; hence it is indeed of the order of 10%, as is usually assumed.

In the case of polar molecules, the orientation polarization must be added to Eq. (4.14):

$$\boldsymbol{P} = N(\langle \boldsymbol{\alpha} \cdot \boldsymbol{E_i} \rangle + \langle \bar{\boldsymbol{\mu}} \rangle), \tag{5.2}$$

where $\bar{\boldsymbol{\mu}}$ is the average value of the dipole moment in the presence of the electric field, and the brackets indicate the average over the orientations of all molecules. Combination of Eq. (5.2) with Eq. (5.1) leads to

$$(\boldsymbol{\varepsilon} - \mathrm{I}) \cdot \boldsymbol{E} = (N/\varepsilon_0)(\langle \boldsymbol{\alpha} \cdot \boldsymbol{E_i} \rangle + \langle \bar{\boldsymbol{\mu}} \rangle). \tag{5.3}$$

Maier and Meier (1961a) have evaluated this expression in the case of nematic liquid crystals, following closely Onsager's theory of the isotropic phase. Therefore we shall first summarize these latter results. For derivations and a more detailed discussion we refer to Böttcher (1973).

To determine the orientation polarization in the isotropic phase, the value of $\bar{\mu}$ is computed from the energy of a dipole in the directing field E_d:

$$U = -\boldsymbol{\mu}\cdot\boldsymbol{E}_d = -\mu E_d \cos\theta, \tag{5.4}$$

where θ is the angle between the directions of $\boldsymbol{E}_d$ and $\boldsymbol{\mu}$. The average value of $\cos\theta$ is determined by the well-known Langevin function. Retaining only the first term in a series expansion this leads to

$$\bar{\mu} = \mu\,\overline{\cos\theta} = \left[\,\mu^2/(3k_{\mathrm{B}}T)\,\right]E_d. \tag{5.5}$$

To describe the permittivity of a condensed phase, we must consider the fields E_i and E_d. According to Onsager the directing field differs from the internal field because of the presence of the reaction field. The dipole polarizes its surroundings, which leads in turn to a reaction field $\boldsymbol{R}$ at the position of the dipole. As $\boldsymbol{R}$ is always parallel to the dipole it cannot direct the dipole, but $\boldsymbol{R}$ does contribute to E_i. The internal field is now given by the directing field E_d, plus the average reaction field $\bar{\boldsymbol{R}}$;

$$E_i = E_d + \bar{\boldsymbol{R}}. \tag{5.6}$$

The reaction field of a dipole moment $\boldsymbol{\mu}$ with polarizability α is given by

$$R = f(\boldsymbol{\mu} + \alpha R) = fF\mu, \tag{5.7}$$

where f is the reaction field factor, which for a spherical cavity with radius a, is:

$$f = (\varepsilon - 1)\left[2\pi\varepsilon_0 a^3(2\varepsilon + 1)\right]^{-1}, \tag{5.8}$$

while

$$F = (1 - f\alpha)^{-1}. \tag{5.9}$$

E_d can be calculated as the sum of the field in a spherical cavity

$$E_c = \left[3\varepsilon/(2\varepsilon + 1)\right]E \equiv hE, \tag{5.10}$$

and the reaction field of the induced dipole moment. This leads to

$$E_d = E_c + f\alpha E_d = FhE. \tag{5.11}$$

With Eqs. (5.5), (5.6) and (5.11) E_i and E_d can be calculated. Substituting the results in the fundamental equation, Eq. (5.3), and replacing α by its average value $\bar{\alpha}$, we arrive at Onsager's result

$$\varepsilon = 1 + (NhF/\varepsilon_0)\left[\bar{\alpha} + F\mu^2/(3k_{\mathrm{B}}T)\right]. \tag{5.12}$$

The factors h and F depend on ε and $\bar{\alpha}$ and on the radius a of the spherical cavity, which is usually taken to be given by

$$\tfrac{4}{3}\pi N a^3 = 1. \tag{5.13}$$

Using this approximation and the value for $\bar{\alpha}$ obtained from the high-frequency permittivity ε_∞ *via* the Clausius-Mossotti equation, Eq. (4.26), we can rewrite Eq. (5.12) as

$$(\varepsilon - \varepsilon_\infty)\frac{2\varepsilon + \varepsilon_\infty}{\varepsilon(\varepsilon_\infty + 2)^2} = \frac{N}{9\varepsilon_0 k_B T}\mu^2. \tag{5.14}$$

With this equation we can compute the value of the permanent dipole moment of a molecule from the dielectric permittivity if the density and ε_∞ are known. Comparison with the actual value of μ as measured for the gas phase provides a check on the validity of the assumptions underlying Eq. (5.14) (see for example Table 16 in Böttcher, 1973). Deviations can be expected, especially when the particles cannot be treated as being spherical, and when specific interactions between the particles occur. Unfortunately, in the event of disagreement it is often difficult to ascertain which factor is responsible.

Maier and Meier (1961a) have extended Onsager's theory to nematic liquid crystals. In their theory a molecule is represented by an anisotropic polarizability $\boldsymbol{\alpha}$ with principal elements α_l and α_t in a spherical cavity of radius a. Furthermore there is a permanent dipole moment $\boldsymbol{\mu}$ that makes an angle β with the direction of α_l. Therefore $\mu_l = \mu\cos\beta$ and $\mu_t = \mu\sin\beta$. In their calculation, they follow Onsager's theory very closely, and this involves the following approximations:

i) In the cavity factor h and the reaction field factor f, the anisotropy of the permittivity is not taken into account. Eqs. (5.10) and (5.8) are used with $\bar{\varepsilon}$ substituted for ε. This approximation is justified as long as $\Delta\varepsilon \ll \bar{\varepsilon}$.

ii) In the calculation of the factor F the anistropy of the polarizability is ignored. Eq. (5.9) is used with $\bar{\alpha}$ substituted for α. This is in fact not correct, as $\alpha_l - \alpha_t$ is of the same order of magnitude as α_t.

These approximations imply that the fields E_i and E_d are taken to be isotropic, just as in the case of Vuks' equation for non-polar molecules. As in that case, we can therefore write for the induced polarization [Eq. (4.14)]:

$$\boldsymbol{P} = N\langle\boldsymbol{\alpha}\rangle\cdot\boldsymbol{E}_i, \tag{5.15}$$

and, as with Eq. (3.19):

$$\langle\alpha\rangle_{\parallel}=\tfrac{1}{3}\left[\alpha_l(2S+1)+\alpha_t(2-2S)\right],$$

$$\langle\alpha\rangle_{\perp}=\tfrac{1}{3}\left[\alpha_l(1-S)+\alpha_t(2+S)\right].\tag{5.16}$$

The calculation of the orientation polarization proceeds as follows. The potential energy U of the dipole moment in the electric field is given by Eq. (5.4); the total energy is $U+W$, where W is the nematic potential. Expanding U, and retaining only terms linear in the electric field, we get:

$$\exp\left[-(U+W)/k_BT\right]\approx(1-U/k_BT)\exp(-W/k_BT).\tag{5.17}$$

The average value of μ_γ ($\gamma=\parallel,\perp$) in the presence of the electric field is then calculated to be

$$\langle\bar{\mu}_\gamma\rangle=\langle\mu_\gamma(1+\mu_\gamma E_d/k_BT)\rangle=\langle\mu_\gamma^2\rangle E_d/k_BT.\tag{5.18}$$

The dipole component along the z axis is given by:

$$\mu_{\parallel}=\mu_l\cos\theta+\mu_t\sin\psi\sin\theta,\tag{5.19}$$

where ψ is the angle between the μ,ζ plane and the normal to the z,ζ plane (compare Figure 1.4). Using this expression for $\mu_{\parallel}$, and noting that $\langle\sin^2\psi\rangle=\tfrac{1}{2}$ because of the assumed axial symmetry of the molecules, we get:

$$\begin{aligned}\langle\mu_{\parallel}^2\rangle&=\mu_l^2\langle\cos^2\theta\rangle+\tfrac{1}{2}\mu_t^2\langle\sin^2\theta\rangle\\&=\tfrac{1}{3}\left[\mu_l^2(2S+1)+\mu_t^2(1-S)\right]\\&=\tfrac{1}{3}\mu^2\left[1-(1-3\cos^2\beta)S\right].\end{aligned}\tag{5.20a}$$

After a similar calculation we find:

$$\begin{aligned}\langle\mu_{\perp}^2\rangle&=\tfrac{1}{3}\left[\mu_l^2(1-S)+\tfrac{1}{2}\mu_t^2(2+S)\right]\\&=\tfrac{1}{3}\mu^2\left[1+\tfrac{1}{2}(1-3\cos^2\beta)S\right].\end{aligned}\tag{5.20b}$$

For $S=1$ only the component μ_l contributes to $\langle\mu_{\parallel}^2\rangle$, and only μ_t contributes to $\langle\mu_{\perp}^2\rangle$.

Inserting the expressions for E_i and E_d as for the isotropic case, Eq. (5.3) can be written as

$$\varepsilon_\gamma=1+(NhF/\varepsilon_0)(\langle\alpha\rangle_\gamma+F\langle\mu_\gamma^2\rangle/k_BT),\qquad\gamma=\parallel,\perp,\tag{5.21}$$

where $\langle\alpha\rangle_\gamma$ is given by Eq. (5.16) and $\langle\mu_\gamma^2\rangle$ by Eq. (5.20). Combination of Eq. (5.21) for $\gamma=\parallel$ and $\gamma=\perp$ leads to an expression for the dielectric anisotropy

$$\Delta\varepsilon = (NhF/\varepsilon_0)\left[\alpha_l - \alpha_t - F(\mu^2/2k_{\mathrm{B}}T)(1-3\cos^2\beta)\right]S. \qquad (5.22)$$

If we take $\mu = 0$, Eq. (5.21) reduces Eq. (4.25). This can be seen by substituting Eq. (5.10) and Eq. (5.9) for h and F, respectively (with ε and α replaced by $\bar{\varepsilon}$ and $\bar{\alpha}$), and using Eq. (5.13). As already noted, the approximations used by Maier and Meier are exactly the same as those when Vuks' isotropic internal field is used. Furthermore, if we calculate $\bar{\varepsilon}$, Onsager's equation (Eq. 5.12) is reproduced with ε replaced by $\bar{\varepsilon}$. Of course, Eq. (5.21) then also reduces to Onsager's equation, in the limit $S = 0$. Hence $\bar{\varepsilon}$ is predicted to be continuous at T_{NI} (apart from the small density change). Contrary to the case of nematics consisting of non-polar molecules, this is not observed experimentally. Finally we note that in the other limit, $S = 1$, the orientation polarization is predicted to contribute still to the permittivity. In practice, rotation of the molecules around a short axis will become more and more difficult when S approaches unity, which means that the relaxation frequency associated with $\varepsilon_{\parallel}$ shifts to lower and lower frequencies (see Section 5.2). This effect concerns the dynamics of the process and does not invalidate the predictions of the static theory.

At first sight, it seems attractive to extend the spheroidal model leading to Eq. (4.29) to polar molecules. However, contrary to the situation for non-polar molecules there is no experimental evidence that E_i and/or E_d are independent of the macroscopic dielectric anisotropy. Consequently specific assumptions have to be made to extend Maier and Meier's theory, and these can hardly be tested. The choice of assumption made by Derzhanski and Petrov (1971) in their spheroidal model cannot be correct as their theory does not predict $\bar{\varepsilon}/\rho$ to be continuous at the NI transition in the limit $\mu = 0$.

Maier and Meier's equations account satisfactorily for many essential features of the permittivity of nematic liquid crystals with polar molecules. This is best illustrated using Eq. (5.22) for $\Delta\varepsilon$. If $3\cos^2\beta = 1$ ($\beta \approx 55°$) the dipole moment contributes equally to $\varepsilon_{\parallel}$ and $\varepsilon_{\perp}$. Then $\Delta\varepsilon$ is determined by the (positive) anisotropy of the polarizability. The dipole contribution to $\Delta\varepsilon$ is positive for $\beta < 55°$ and negative for $\beta > 55°$. In the latter case, whether $\Delta\varepsilon$ itself becomes negative depends on the relative magnitude of the two contributions. The contribution of the induced polarization to $\Delta\varepsilon$ varies with temperature like S; the temperature dependence of the orientation polarization varies with S/T. The temperature dependence of the other factors is weak. This explains nicely the following observations:

i) When $\Delta\varepsilon$ is negative or strongly positive, the S/T dependence of the dipole contribution to $\Delta\varepsilon$ is predominant over the whole temperature range. Consequently $|\Delta\varepsilon|$ increases with decreasing temperature (Figures 5.1b, 5.2a and 5.2b).

ii) When $\Delta\varepsilon$ is positive and close to zero, the anisotropy of the induced polarization approximately equals that of the orientation polarization. Just below T_{NI} the temperature dependence of S is most important and $\Delta\varepsilon$ increases with decreasing temperature. At lower temperatures the variations of S are small and the counteracting S/T dependence of the orientation polarization predominates. Then we may find that $\Delta\varepsilon$ first increases with decreasing temperature, goes through a maximum, and finally decreases with decreasing temperature. This effect is illustrated in Figure 5.3b.

It is difficult to test Eqs. (5.21) and (5.22) more quantitatively. For PAA, information about $\bar{\alpha}$ and $\alpha_l - \alpha_t$ is available from optical measurements. The Onsager equation can then be used to calculate $\mu = 7.4 \times 10^{-30}$ Cm from ε for the isotropic phase,[†] and then Eq. (5.22) to calculate $\beta = 65°$ from $\Delta\varepsilon$. The result for μ agrees reasonably well with the value of 7.7×10^{-30} Cm obtained from dilute solutions of PAA in benzene (Maier and Meier, 1961b). In principle the results for μ and β can be checked with the results from measurements of the Kerr constant, and reasonable agreement is obtained. As already discussed on page 45, this does not necessarily justify the approximations.

As an example of how Eqs. (5.21) and (5.22) can be used, at least qualitatively, we shall discuss the influence of terminal groups on the permittivities of p,p'-disubstituted azobenzenes. In Figure 5.1a, the permittivities of p,p'-diheptylazobenzene are given. As this compound is nonpolar, it can act as a "reference" system. In Figure 5.3, the permittivities are given for some terminally substituted azobenzenes in which the end groups have been varied systematically. We shall first compare Figure 5.1a with Figure 5.3a. Replacing one heptyl group by an alkoxy group of similar size leads to a small increase of both $\varepsilon_\parallel$ and $\varepsilon_\perp$. At a reduced temperature of $0.97T_{\mathrm{NI}}$, $\Delta\varepsilon$ has changed from a value of 0.4 to 0.3. The group moment of an $-OCH_3$ group is 5.6×10^{-30} Cm at $72°$ to the *para* axis of the adjacent aromatic ring (Minkin *et al.* 1970). We shall assume that this is also a typical value for alkyl chains longer than $-CH_3$. The longitudinal component of the alkoxy dipole moment counteracts the dipole of the heptyl group (about 1.3×10^{-30} Cm along the *para* axis) at the other *para* position. However, exact cancellation, as in the case of dialkyl substitution, cannot be expected. Furthermore, the polarizability along the *para* axis will be somewhat greater than in the case of dialkyl

---

[†]1 Cm equals 0.30×10^{30} D.

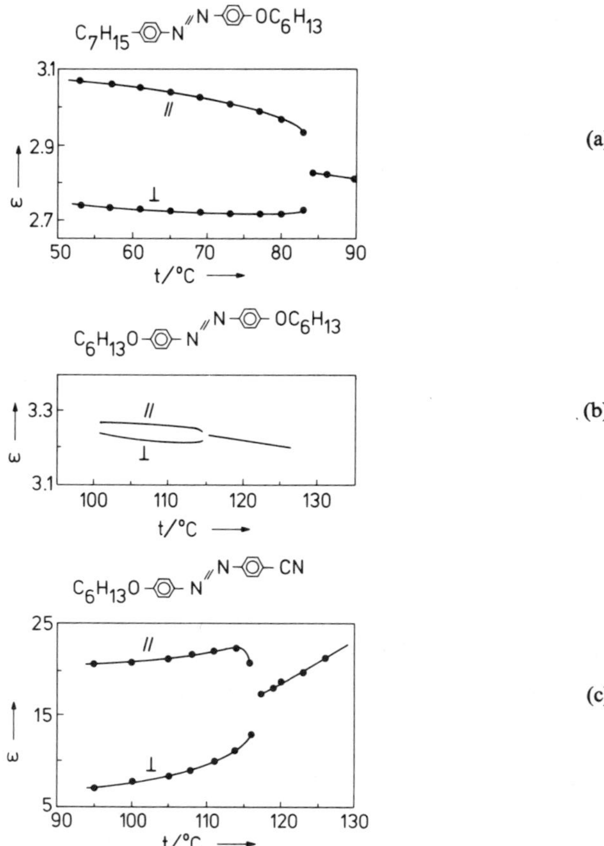

FIGURE 5.3   Dielectric permittivities of some terminally substituted azobenzenes (de Jeu, 1978a).

substitution. These two effects explain the increase of $\varepsilon_\parallel$. The increase of $\varepsilon_\perp$ is due to the transverse component of the alkoxy dipole moment.

When two alkoxy groups are introduced (Figure 5.3b) the transverse component of the total dipole moment is approximately a factor of $\sqrt{2}$ larger than that in the case of Figure 5.3a. (The dipole moments add up quadratically, when free rotation of the alkoxy groups is assumed). However, as $\varepsilon_\parallel$ has also increased somewhat due to the greater polarizability along the *para* axis, $\Delta\varepsilon$ is still (small) positive. Finally we consider the case of cyano substitution (Figure 5.3c). The cyano group has a large dipole moment of 13.5×10^{-30} Cm along the *para* axis. The contribution to the permittivity of this dipole moment dominates over all other contributions. As the angle between the *para* axis and the molecular ζ axis is small, this leads to a large value of $\varepsilon_{\parallel}$ and consequently to a large positive $\Delta\varepsilon$. Note that $\bar{\varepsilon}$ in Figure 5.3c decreases with decreasing temperature, contrary to the prediction from Eq. (5.12) and the usual trend. This effect is typical of compounds that associate with their dipole moments anti-parallel, the association becoming stronger at lower temperatures.

Considerations of this kind can be quite useful in predicting roughly the dielectric anisotropy of a compound from the group moments. A comparison of various Schiff's bases and phenyl benzoates has been given by Klingbiel *et al.* (1974b), while various bridging groups have been compared by de Jeu and Lathouwers (1975). Other papers dealing with molecular structure and dielectric permittivity are Maier and Meier (1961c); Diguet *et al.* (1970); Schadt (1972); Klingbiel *et al.* (1974a), and Kresse *et al.* (1975, 1977).

The notable disagreement between Maier and Meier's theory and experiments is that $\bar{\varepsilon}$ for the nematic phase, when extrapolated to T_{NI}, is always lower than the isotropic value of ε at T_{NI}. The difference will be denoted here by $\delta\varepsilon_{\mathrm{NI}}$. Investigation of this effect in various compounds leads to the conclusion that $\delta\varepsilon_{\mathrm{NI}}$ is independent of the sign and magnitude of $\Delta\varepsilon$ (de Jeu, 1978a). In principle, this effect could be attributed either to the omission of the various anisotropies in the calculation of E_i and E_d, or to a systematic change in specific intermolecular interactions at T_{NI}. As $\Delta\varepsilon$ has little or no influence on $\delta\varepsilon_{\mathrm{NI}}$, the latter assumption seems to be more plausible. We shall meet some examples of this effect at the end of this section.

In the Onsager theory, the dielectric properties of polar liquids are investigated by considering a molecule in the average field of the other molecules. The environment of a molecule is treated as a continuous medium. In this way specific short-range interactions are not taken into account. In the case of isotropic polar liquids, the Kirkwood-Fröhlich theory provides, in principle, a framework in which such interactions can be explicitly evaluated (Böttcher, 1973, Ch. VI; Bordewijk, 1973). In their treatment, the number of molecules treated statistically is restricted by taking a small, but still macroscopic sphere with volume v, imbedded in an infinite dielectric with permittivity ε. The permittivity can then be related to the total electric moment of the sphere. Calculation of the average value of this moment lead to

$$(\varepsilon - \varepsilon_\infty)\frac{2\varepsilon + \varepsilon_\infty}{\varepsilon(\varepsilon_\infty + 2)^2} = (9\varepsilon_0 k_{\mathrm{B}} T)^{-1}\left\langle \sum_i \sum_j \boldsymbol{\mu}_i \cdot \boldsymbol{\mu}_j \right\rangle$$

$$= \left[N/(9\varepsilon_0 k_{\mathrm{B}} T) \right] g\mu^2, \tag{5.23}$$

where the summation extends over all molecules in the sphere, and g is the Kirkwood correlation factor. The factors connected with $\varepsilon - \varepsilon_\infty$ are related to the polarization of a sphere with permittivity ε_∞ surrounded by a dielectric with permittivity ε, and to the cavity field in this sphere. Contrary to the situation in Onsager's theory, these expressions have a precise meaning as the sphere under consideration is macroscopic. In the absence of correlation, $g = 1$ and Eq. (5.23) reduces to Onsager's equation, Eq. (5.14). In order to calculate g in real cases, specific assumptions must be made about the interaction in the systems considered.

In order to extend Eq. (5.23) to liquid crystals, we must consider a macroscopic sphere in a uniaxial anisotropic dielectric with permittivities $\varepsilon_\parallel$ and $\varepsilon_\perp$. In this situation the cavity field and related problems can be calculated as in the isotropic case (Böttcher and Bordewijk, 1978, Sec. 94) if the following transformation is made:

$$x' = x\varepsilon_\perp^{-1/2}, \qquad y' = y\varepsilon_\perp^{-1/2}, \qquad z' = z\varepsilon_\parallel^{-1/2}. \tag{5.24}$$

In these new coordinates the sphere has changed to a spheroid with axial ratio $(\varepsilon_\parallel / \varepsilon_\perp)^{1/2}$, and shape factors enter into the picture that depend only on the anisotropy of the permittivity. In order to distinguish them from those used previously for molecular spheroids, we shall denote these factors by $\Omega_\gamma^\varepsilon$ ($\gamma = \parallel, \perp$). If we use Vuks' equation for the internal field in the calculation of the moment of the sphere, the anisotropic Kirkwood-Fröhlich equation reads (Bordewijk, 1974)

$$(\varepsilon_\gamma - \varepsilon_{\infty\gamma}) \frac{\varepsilon_\gamma + (\varepsilon_{\infty\gamma} - \varepsilon_\gamma)\Omega_\gamma^\varepsilon}{\varepsilon_\gamma(\bar{\varepsilon}_\infty + 2)^2} = (9\varepsilon_0 k_B T v)^{-1} \left\langle \sum_i \sum_i (\mu_i)_\gamma \cdot (\mu_j)_\gamma \right\rangle$$

$$= \left[N/(9\varepsilon_0 k_B T) \right] g_\gamma \langle \mu_\gamma^2 \rangle, \qquad \gamma = \parallel, \perp, \tag{5.25}$$

where $\langle \mu_\gamma^2 \rangle$ is given by Eq. (5.20). If the spheroidal model for the internal field is used, slightly different equations are obtained (Bordewijk and de Jeu, 1978). We shall use Eq. (5.25) as the basis of the discussion of two rather different examples of correlation effects.

(a)   *The influence of smectic order*

Figure 5.4 gives the static permittivity of p,p'-diheptylazoxybenzene in the isotropic, the nematic and the smectic A phases. Just below T_{NI}, $\Delta\varepsilon$ increases as usual with decreasing temperature. However, at lower temperatures, this trend is reversed, leading finally to a change of sign of $\Delta\varepsilon$. A comparison with other compounds of the series reveals that this effect is related to the occurrence of the smectic phase (de Jeu *et al.* 1974). In this

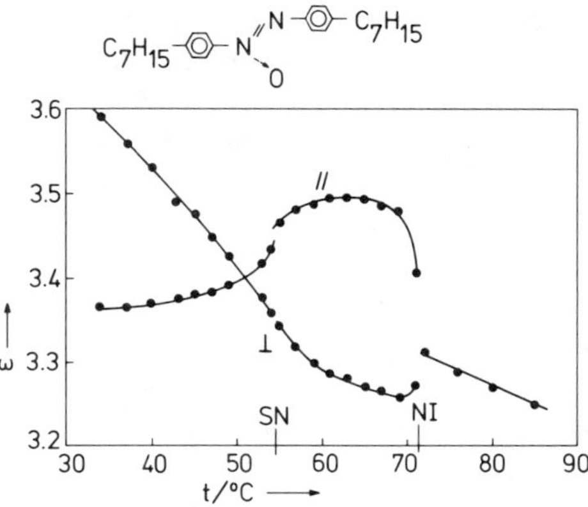

FIGURE 5.4   Dielectric permittivities of p,p'-diheptylazoxybenzene (de Jeu, 1978a).

phase, the interaction of a dipole moment with the dipoles of surrounding
molecules works out differently from that in the nematic phase, owing to
the non-isotropic distribution of the centres of mass. For dipoles situated
in the central part of the molecules, the distance between the dipoles of
molecules in different smectic layers is much greater than the distance
between neighbouring dipoles in the same layer. For the dipole components along the director this leads to an increased anti-parallel correlation
(see Figure 5.5). Consequently, the effective moment in this direction is
reduced, leading to a decrease of $\varepsilon_{\parallel}$. An increase of $\varepsilon_{\perp}$ can be similarly
explained. Note that because the average value of a dipole μ in an electric
field is proportional to μ^2, see Eq. (5.5), rotation of the molecules around a
long or short axis does not affect this result. Because the nematic-smectic
A transition of p,p'-diheptylazoxybenzene is almost second-order (only a

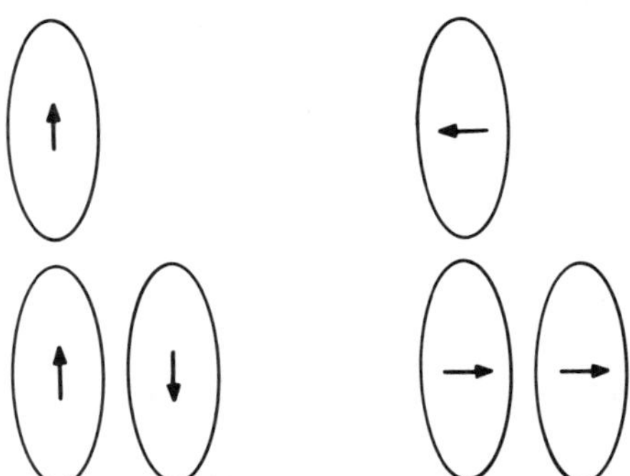

FIGURE 5.5   Dipole correlation between neighbouring molecules.

very small transition heat is observed), large effects due to pre-transitional smectic order are already present in the nematic phase.

In this situation, the right-hand side of Eq. (5.25) can be calculated as follows. Consider a fixed molecule of spheroidal shape with long axis $2a$ and short axis $2b$, in a macroscopic sphere. The average moment of the sphere in the field of the central dipole is given by $\bar{\mu}$. We restrict ourselves to the component $\bar{\mu}_{\parallel}$, which can be calculated in a very approximate way by taking ideal order ($S=1$) and restricting the radius of the sphere to a. The average moment of a particle at a distance r from the fixed molecule is taken as $\mu_{\parallel}^2 E(r)/k_B T$, where $E(r)$ is the field of the central dipole [compare Eq. (5.5)]. Then in the nematic case

$$\bar{\mu}_{\parallel}(\text{nem}) = \mu_{\parallel} + \left(N\mu_{\parallel}^2/k_B T\right)\int dV\, E, \qquad (5.26)$$

where the integration is over the volume of the sphere outside the molecular spheroid, and $N = 3/(4\pi ab^2)$ is the dipole density. The integral is given by

$$\int dV\, E = -\Omega_{\parallel}\,\mu_{\parallel}/\varepsilon_0, \qquad (5.27)$$

where the shape factor $\Omega_{\parallel}$, given by Eq. (4.12), is equal to $\frac{1}{3}$ for the sphere, while for a spheroid with $a \gg b$ we have $\Omega_{\parallel} \ll \frac{1}{3}$. Then

$$\bar{\mu}_{\parallel}(\text{nem}) \approx \mu_{\parallel}\left[1 - \left(\mu_{\parallel}^2/k_B T\right)(4\pi\varepsilon_0 ab^2)^{-1}\right]. \qquad (5.28)$$

For the smectic A phase, we assume, in addition to $S=1$, that the centres of mass of the molecules are exactly arranged in layers. The integration over the volume of the sphere then reduces to an integration over a circular surface A:

$$\bar{\mu}_{\parallel}(\text{sm}) = \mu_{\parallel} + \left(N'\mu_{\parallel}^2/k_B T\right)\int da\, E, \qquad (5.29)$$

where now $N' = 1/(\pi b^2)$ is the surface dipole density. Over the surface

$$\int dA\, E = (\mu_{\parallel}/2\varepsilon_0)(1/a - 1/b). \qquad (5.30)$$

Again using $a \gg b$ we find

$$\bar{\mu}_{\parallel}(\text{sm}) = \mu_{\parallel}\left[1 - \left(\mu_{\parallel}^2/k_B T\right)(2\pi\varepsilon_0 b^3)^{-1}\right]. \qquad (5.31)$$

Comparing Eqs. (5.28) and (5.31) we see that for the smectic phase the correction to $\mu_{\parallel}$ is a factor of $2a/b \approx 10$ larger than that for the nematic phase. In fact, the corrections thus calculated are much too large, which is not surprising in the light of the approximations made. Nevertheless, the calculations indicate that $g_\parallel < 1$, while $g_\parallel$ is much smaller in the smectic phase than in the nematic phase, in agreement with the experiments. One can calculate $g_\perp > 1$ in a similar way.

(b)   *Association in para-cyano substituted compounds*

In Figure 5.2b, we presented the permittivity of p-heptyl-p'-cyanobiphenyl. Due to the relatively large dipole moment of the cyano group parallel to the long molecular axis, a large positive dielectric anisotropy is observed for the nematic phase. Furthermore, $\bar{\varepsilon}$ for the nematic phase and ε for the isotropic phase increase with increasing temperature. Calculation of the correlation factors gives a value of $g_\parallel$ which is much smaller than one, indicating strong anti-parallel correlation of the dipole moments. It seems that the association in these strongly polar substances has a profound effect on the properties of the liquid crystalline phase. By way of example we mention the following peculiarities:

i) p,p'-substituted biphenyls give smectic phases for almost all the usual end substituents. It is only when a strongly polar end group ($-CN$, $-NO_2$, *etc.*) is present that nematic phases also occur (Schubert and Dehne, 1972; Gray, 1975).

ii) When mesogenic compounds with a strongly polar end group give a smectic phase, the layer thickness is usually about 1.4 times the length of the all-stretched molecule. Hence some type of double layer formation occurs (see, for example, Leadbetter *et al.* 1977).

iii) Binary mixtures of a mesogenic compound with a strongly polar end group and one in which no such group is present give very peculiar phase diagrams. Often a smectic phase is stabilized strongly around the 1:1 composition range, even if the pure compounds do not show smectic behaviour at all (Oh, 1977).

iv) If one or both of the phenyl rings in the p,p'-substituted biphenyl is/are replaced by a cyclohexane ring, nematic behaviour is usually still observed, but only if a strongly polar end group is present (Eidenschink *et al.* 1978).

All these effects are to a large extend unexpected on the basis of the knowledge obtained with more traditional types of mesogenic compound. Obviously the amount of association due to the strongly polar groups is a determining factor, and dielectric methods will probably prove to be a very useful means of obtaining a better understanding of these effects.

Other examples of interesting correlation effects are observed with the p-substituted benzoic acids, which exist largely as dimers (Kresse *et al.* 1977). All these examples show that dipole correlation can be quite important. The observed values of $\delta\varepsilon_{NI}$ are probably attributable to a discontinuous change in the correlation factors at T_{NI}, the antiparallel correlation being somewhat stronger in the nematic phase.

5.2 DIELECTRIC RELAXATION

So far we have restricted ourselves to the permittivity in static fields. After the field is removed the orientation polarization decays exponentially with a characteristic time τ, the relaxation time. The process of (re)orientation of the permanent dipole moments connected with changes in the field requires a definite time interval. In alternating fields this leads to a time lag between the average orientation of the dipole moments and the field. This effect becomes noticeable at frequencies of the order of τ^{-1}. At much higher frequencies the orientation polarization can no longer follow the variation of the field. The residual permittivity ε_{∞} is due to the induced polarization only. In liquid crystals we meet the situation where the relaxation occurs with an anisotropic permittivity. The difference in relaxation behaviour of $\varepsilon_{\parallel}$ and $\varepsilon_{\perp}$ can, in principle, provide information about the dynamic behaviour of the molecules in the liquid crystalline phase.

To measure the permittivity up to frequencies of about 10 MHz, standard capacitance bridges can still be used, and the same measurement configuration applies as at low frequencies. Nevertheless care must be taken to account for possible effects of the series resistance and inductance of the measurement leads and electrodes, which are incorporated in the effectively measured parallel capacitance and resistance. Transparent indium-oxide and tin-oxide electrodes can have an appreciable resistance, the effect of which may become noticeable even below 1 MHz. Above 10 MHz, a lumped element picture of the measurement set-up becomes gradually meaningless. Resonant circuits are often used (see, for example, Vaughan, 1969). A measurement cell for liquid crystals suitable for the region above 10 MHz, built upon a coaxial line, has been described by Druon and Wacrenier (1977). Measurements of the permittivity of liquid crystals above 10 MHz are relatively rare. This is disappointing, as the time scales of these high frequency relaxations would nicely complement those of neutron scattering and nuclear magnetic resonance.

For isotropic liquids, the relaxation of the permittivity can be described by a complex generalized permittivity $\varepsilon^* = \varepsilon' - i\varepsilon''$. In the case of an alternating field of circular frequency ω Debye's classical theory (see, for example, Hill, 1969) leads to

$$\varepsilon^*(\omega) = \varepsilon_\infty + (\varepsilon - \varepsilon_\infty)/(1 + i\omega\tau), \qquad (5.32)$$

where ε and ε_∞ are the static and the high-frequency permittivity respectively. If $\omega\tau$ is eliminated from the real and imaginary parts of Eq. (5.32) the equation of a circle is obtained. Consequently a plot of ε'' vs. ε' should give a semicircle if the relaxation can be described by Eq. (5.32). In the case of a liquid crystal this equation can be applied to each of the components $\varepsilon_\parallel^*$ and $\varepsilon_\perp^*$. Figure 5.6 shows such so-called Cole-Cole plots for

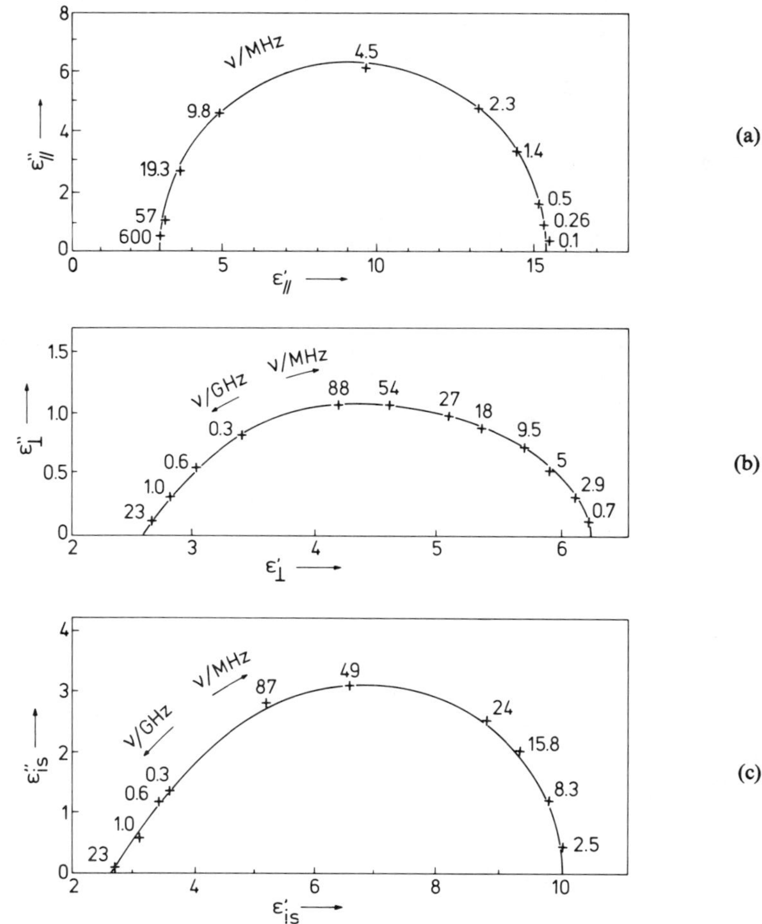

FIGURE 5.6 Cole-Cole plots for the dielectric relaxation of p-heptyl-p'-cyanobiphenyl (Lippens et $al.$ 1977).

p-heptyl-p'-cyanobiphenyl. For $\varepsilon_\parallel^*$ a semicircle is indeed obtained; in the other cases deviations are observed, indicating a distribution of relaxation times. In Figure 5.7 the frequency of maximum loss (maximum ε'') is plotted against $1/T$ for the same compound.

Eq. (5.32) was originally derived by Debye, who assumed that the tendency of the molecular dipole moments to align parallel to the applied field is counteracted by rotational diffusion. Denoting the angle between the dipole moment and the field by θ, the potential energy of a molecule U is given by Eq. (5.4), and the directing torque is $\Gamma = -\partial U/\partial\theta$. Let $f(\theta,t)d\Omega$ be the fraction of molecules with their dipole moment in the direction $d\Omega = 2\pi \sin\theta\, d\theta$. Taking the diffusive process into account, Debye's equation for $f(\theta,t)$ is obtained as:

$$\frac{\partial f(\theta,t)}{\partial t} = \frac{D}{\sin\theta}\frac{\partial}{\partial\theta}\left\{\sin\theta\left[\frac{\partial f(\theta,t)}{\partial\theta} - \frac{\Gamma(\theta)}{k_B T}f(\theta,t)\right]\right\}, \qquad (5.33)$$

where D is the rotational diffusion constant. If a field E_0 has been switched on at $t = -\infty$, we have at $t = 0$

$$f(\theta,0) = A\exp(-U/k_B T)$$
$$\approx A\left[1 + (\mu E_0/k_B T)\cos\theta\right], \qquad (5.34)$$

where A is a constant. Hence in the case of an alternating field $E = E_0\exp(i\omega t)$ we may try solutions of the form

$$f(\theta,t) = A\left[1 + B(t)(\mu E_0/k_B T)\cos\theta\right]. \qquad (5.35)$$

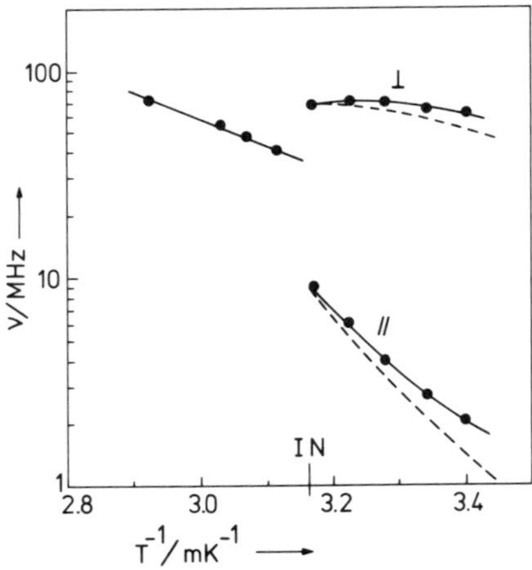

FIGURE 5.7   Relaxation frequencies $\nu = 1/2\pi\tau$ for p-heptyl-p'-cyanobiphenyl. Filled dots: experimental points (Davies *et al.* 1976); broken line: calculated theoretically (see text).

Substitution in Eq. (5.33) gives

$$B = (1 + i\omega\tau')^{-1}, \qquad \tau' = 1/2D. \tag{5.36}$$

From the result for $f(\theta, t)$ we can calculate the orientation polarization, and its relation with the complex permittivity. With the low-frequency limit (giving ε) and the high-frequency limit (giving ε_∞) Eq. (5.32) is obtained. The relation between τ and τ' depends on the particular model of the internal field.

In order to understand what happens in a nematic liquid crystal, we have to consider in detail the contributions of μ_l and μ_t to $\varepsilon_\parallel$ and $\varepsilon_\perp$, respectively (see Figure 5.8). We have:

$$\varepsilon_\parallel - \varepsilon_{\infty\parallel} \sim \langle \mu_\parallel^2 \rangle = \mu_l^2 \langle \cos^2\theta \rangle + \tfrac{1}{2}\mu_t^2 \langle \sin^2\theta \rangle,$$

$$\varepsilon_\perp - \varepsilon_{\infty\perp} \sim \langle \mu_\perp^2 \rangle = \mu_l^2 \langle \sin^2\theta \rangle + \tfrac{1}{2}\mu_t^2 \langle \cos^2\theta \rangle, \tag{5.37}$$

where θ is now the angle between the direction of μ_l and the field. A reorientation of μ_t can be accomplished by a rotation around the long molecular axis, which, to first order in E, is independent of the nematic order. For $\varepsilon_\perp$ this is the main contribution to the relaxation process, which is thus not influenced by the nematic potential. For $\varepsilon_\parallel$ the contribution of μ_t is in general less important, and moreover it decreases with increasing value of S. The interesting differences compared with the relaxation in the isotropic phase are related to the reorientation of μ_l. For the field along the director ($\varepsilon_\parallel$), a reorientation of μ_l must be accomplished by a rotation around a short molecular axis. Hence this effect is counteracted by the nematic potential, leading to relatively large values of τ (low relaxation frequencies). This will be easily observed, as $\langle \cos^2\theta \rangle$ is also large at large

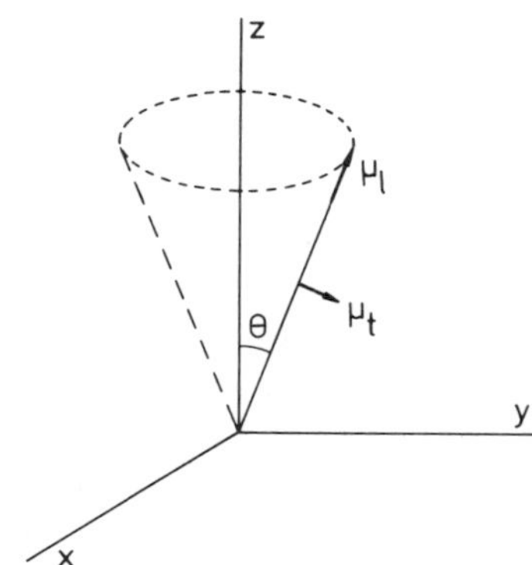

FIGURE 5.8   Reorientation of the dipole component $\mu_\perp$.

values of S, and this is the main contribution to $\varepsilon_{\parallel}$. In the case of a field perpendicular to the director ($\varepsilon_{\perp}$), a reorientation of μ_l is accomplished by rotation through an angle π over the latitude $\theta = $ constant (see Figure 5.8). The smaller the angle θ, the easier this is. Hence the nematic potential now leads to a decrease of the relaxation time (increase of the relaxation frequency). In general this will be difficult to observe as $\langle \sin^2\theta \rangle$ is small at large values of S, and the main contribution to $\varepsilon_{\perp}$, which comes from μ_t, is unaffected.

Relatively low relaxation frequencies for $\varepsilon_{\parallel}$ in the nematic phase have been observed by various authors (Maier and Meier, 1961d; Axmann, 1968; Rondelez and Mircéa-Roussel, 1974; Agarwal and Price, 1974; de Jeu and Lathouwers, 1974b; Bata and Molnar, 1975). The most suitable compounds for the observation of these effects are those with $\mu = \mu_l$ and $\mu_t = 0$. Figure 5.7 shows the relaxation frequencies for p-heptyl-p'-cyano-biphenyl, a compound satisfying these requirements. When the temperature is lowered to below T_{NI}, shifts of the relaxation frequency are indeed observed in the expected directions. The whole argument can be made somewhat more quantitative by assuming a mean-field nematic potential of the form (compare Eq. 1.5)

$$W = -q\cos^2\theta. \tag{5.38}$$

Then an additional torque $\Gamma_{\text{N}} = 2q\sin\theta\cos\theta$ must be added to the electric torque $\Gamma(\theta)$ in Eq. (5.33). The distribution function can be written as $f(\theta,t) = F(\theta,t)w(\theta)$, where $w(\theta)$ is the nematic distribution function in the absence of the electric field. Thus a differential equation for $F(\theta,t)$ is obtained, which leads to a fairly complicated eigenvalue problem, that has been solved by Martin, Meier and Saupe (1971). The result can be expressed as a shift compared with the relaxation time for $q = 0$ (no nematic potential):

$$\tau_\gamma(q) = g_\gamma \tau_\gamma(q=0), \qquad \gamma = \parallel, \perp \tag{5.39}$$

The g_γ are retardation factors (not to be confused with the correlation factors) that depend on q/kT, the strength of the nematic potential (see Figure 5.9). Using Maier and Saupe's molecular-statistical theory of the nematic phase (see Chapter 1), the retardation factors can be calculated for various values of q/kT or S, as a function of the reduced temperature $TV^2/(T_{\text{NI}}V_{\text{NI}}^2)$ (see Table 5.1). At the NI transition we find $g_{\parallel} = 4.0$ and $g_{\perp} = 0.52$, in excellent agreement with the experimental values 4.2 and 0.55 from Figure 5.7. The broken line in Figure 5.7 has been calculated from the extrapolated isotropic relaxation frequency, the values of $g_{\parallel}$ and $g_{\perp}$ from Table 5.1, and the densities as given by Dunmur and Miller (1979).

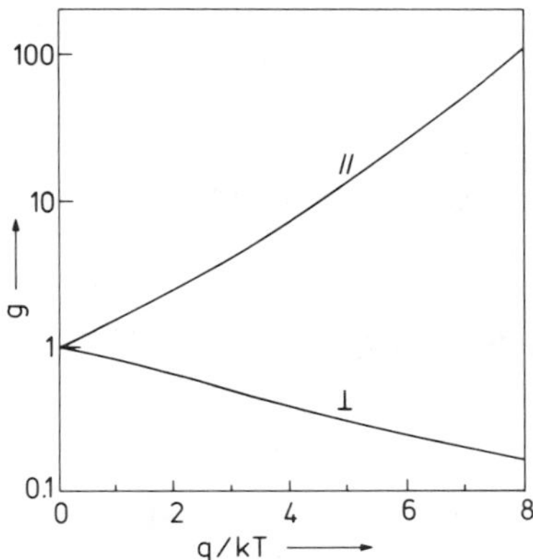

FIGURE 5.9 Retardation factors as a function of the strength of the nematic potential (Martin *et al.* 1971).

TABLE 5.1
Retardation factors calculated using Maier and Saupe's mean-field nematic potential

$TV^2/(T_{NI}V_{NI}^2)$	S	$g_\parallel$	$g_\perp$
1.000	0.429	4.0	0.52
0.985	0.479	5.0	0.48
0.970	0.516	6.0	0.43
0.941	0.569	8.2	0.39
0.912	0.610	11	0.37
0.884	0.643	15	0.35

Looking in some more detail at the experimental data relating to Figure 5.6, we see that for $\varepsilon_\parallel$ there is indeed a single macroscopic relaxation time, but that the relaxation of $\varepsilon_\perp$ is strongly distributed. The Cole-Cole plot for $\varepsilon_\perp$ can be divided into three Debye-type relaxations. The most important one can be identified with the retarded relaxation described above. It is accompanied by a relaxation at lower frequencies (of the same order of magnitude as the one observed for $\varepsilon_\parallel$) and one at higher frequencies. There is no certainty yet regarding the mechanisms involved.

Various attempts have been made to calculate the quantity D, which has been related to viscosity coefficients (see, for example, Hill, 1969). In extrapolating the isotropic relaxation frequency into the nematic phase, we have assumed that: (i) D does not show any discontinuity at T_{NI}, and (ii) the temperature dependence of D is the same in the isotropic and in the nematic phase. This would mean that D depends on the local short-range structure of the liquid and is independent of the nematic potential. From the almost perfect agreement between theoretical and experimental values of $g_\parallel$ and $g_\perp$ at T_{NI}, one could conclude that assumption (i) is correct. The increasing discrepancy between calculated and experimental values at lower temperatures could be attributed to point (ii).

Other experimental results on compounds that have $\mu = \mu_l$ and $\mu_t = 0$ are not yet available. If $\mu_t \neq 0$ the interpretation of the relaxation is much more complicated. In that case $\tau(q=0)$ can no longer be identified with τ_{is}, which makes it difficult to determine the retardation factors (Parneix *et al.* 1975; Mircéa-Roussel and Rondelez, 1975; Schadt, 1972). If τ_{is} is nevertheless used to calculate $g_{\parallel}$, as has been done for PAA (Meier and Saupe, 1966), incorrect values for $g_{\parallel}$ are obtained that are much larger than those expected theoretically.

In earlier discussions of dielectric relaxation in nematics the explicit solution of the differential equation for the distribution function was circumvented by assuming $F(\theta,t) \approx F(\theta,0) = 1 + a(E,t)\cos\theta$. The result for $g_{\parallel}$ is (Meier and Saupe, 1966)

$$g_{\parallel} = (kT/q)\left[\exp(q/k_B T) - 1\right]. \qquad (5.40)$$

Usually $\exp(q/k_B T) \gg 1$, and the last term can be disregarded. The values of $g_{\parallel}$ calculated from Eq. (5.40) are approximately a factor of two higher than those from Figure 5.9. For the isotropic phase, it is often found that the relaxation time $\tau(q=0)$ is proportional to the viscosity η of the liquid. As the temperature dependence of the viscosity can be described by $\eta \sim \exp(q_{\mathrm{visc}}/k_B T)$, combination with Eq. (5.40) leads to

$$\tau_{\parallel}(q) \sim \exp\left[(q + q_{\mathrm{visc}})/k_B T\right]. \qquad (5.41)$$

Although it is *a priori* not clear which of the nematic viscosities should be used (see Chapter 7), Eq. (5.41) has been shown to be obeyed in several experiments, leading to values of $q \approx 30\text{–}50$ kJmol^{-1} (de Jeu and Lathouwers, 1974). The low relaxation frequencies in the kHz region found in some cases are due to a combination of a high viscosity [reducing $\tau(q=0)$] and a low reduced temperature T/T_{NI} (increasing q and thus $g_{\parallel}$). The case where $\Delta\varepsilon$ is initially positive is especially interesting. The relaxation of $\varepsilon_{\parallel}$ can then lead to a reversal of the sign of $\Delta\varepsilon$ in the experimentally easily accessible kHz region. This offers the possibility of studying the behaviour of the nematic in external electric fields as a function of the dielectric anisotropy (de Jeu *et al.* 1972).

We conclude that the extension of Debye's rotational diffusion equation to nematics works remarkably well. Probably this equation is appropriate for compounds forming liquid crystals, because the molecular shape permits neither free rotation nor reorientation by instantaneous jumps. A theory incorporating these other models of rotational motion has been given by Luckhurst and Zannoni (1975). Furthermore, as the short-range order in the isotropic phase and in the nematic phase does not differ much, all conceivable types of corrections will to a first approximation be the same for both phases. The different relaxation in the two phases must then again be attributed to the nematic potential. Therefore one can expect that, in spite of the limitations of the Debye model, the calculation of the retardation factors is fairly reliable. In order to calculate absolute values of $\tau_\parallel$ and $\tau_\perp$, values for D as well as for the internal field have to be calculated from models, as would also be the case if one tried to calculate τ for the isotropic phase.

CHAPTER 6

The elastic constants

Up to now we have essentially considered an ideal single crystal of a nematic. However, if, for example in a thin layer, the orientations of n imposed at the boundaries are not parallel, some curved transition from one orientation to the other is required. Curvature of n is also introduced if the orienting effect of an external field competes with the orientations imposed by the boundaries. These modifications of n take place over macroscopic distances (typically a few microns), and are easily observed optically. On the other hand spatial variations of the order parameter S decrease rapidly (typically over a few molecular lengths). Thus we can say that in a weakly distorted nematic, *i.e.* where the spatial variations of n are small on a molecular scale, the nematic may be pictured as being locally still uniaxial, with a fixed order parameter $S(T)$, but with a variable preferred direction $n(r)$. The complete state of alignment can be expressed in a tensor order parameter $\mathbf{Q}$, with elements

$$Q_{\alpha\beta} = S(T)\left[n_\alpha(r)n_\beta(r) - \tfrac{1}{3}\delta_{\alpha\beta}\right], \qquad \alpha,\beta = x,y,z, \tag{6.1}$$

where $\delta_{\alpha\beta}$ is the Kronecker delta, which has the value unity for $\alpha = \beta$ and is otherwise zero. For a uniform sample with n along the z axis, only the diagonal elements of $\mathbf{Q}$ are non-zero, with

$$Q_{zz} = \tfrac{2}{3}S, \qquad Q_{xx} = Q_{yy} = -\tfrac{1}{3}S.$$

From the theoretical point of view, the curvature of n can be described in terms of a continuum theory, the analogue of the classical elastic theory of a solid. In the latter theory, the material undergoes homogeneous strains, and restoring forces are considered to oppose the change in distance between neighbouring points in the material. In a nematic liquid, where permanent forces opposing the change of distance between points do not exist, we look for restoring torques which directly oppose the *curvature*. Frank (1958) refers to these as torque stresses, and assumes an equivalent of Hooke's law, making them proportional to curvature strains, appropriately defined, when these are sufficiently small. This is equivalent to assuming that the free-energy density is a quadratic function of the curvature strains, in which the analogues of elastic moduli appear as coefficients. Closely following Frank (1958), we shall adopt the latter procedure.

We consider a uniaxial liquid crystal; at each point r the direction of preferred orientation is given by $n(r)$. Once n is defined at some point, we assume that it varies slowly with position, and is defined at other points by continuity. The sign of n is without physical significance. At each point r we introduce a local right-handed Cartesian coordinate system x,y,z, with the z axis along n. In this system, the six linear components of curvature are (see Figure 6.1)

$$s_1 = \partial n_x/\partial x, \qquad s_2 = \partial n_y/\partial y, \qquad \text{(splays)};$$

$$t_1 = -\partial n_y/\partial x, \qquad t_2 = \partial n_x/\partial y, \qquad \text{(twists)}; \qquad (6.2)$$

$$b_1 = \partial n_x/\partial z, \qquad b_2 = \partial n_y/\partial z, \qquad \text{(bends)}.$$

For simplicity we shall restrict ourselves in the following discussion to a planar structure, in which the director remains always perpendicular to a space-fixed axis, say the y axis. Thus we have $\partial n_y = 0$, and only s_1, t_2 and b_1 are possible deformations. Assuming that the free-energy density is a quadratic function of these curvature strains, it is relevant to consider also second-order terms (Nehring and Saupe, 1971). The possibilities are

$$ss = \partial^2 n_x/\partial x^2, \qquad tt = \partial^2 n_x/\partial y^2, \qquad bb = \partial^2 n_x/\partial z^2,$$

$$st = \partial^2 n_x/\partial x\partial y, \qquad sb = \partial^2 n_x/\partial x\partial z, \qquad tb = \partial^2 n_x/\partial y\partial z. \qquad (6.3)$$

Terms linear in these second derivatives of n are of the same order of magnitude as quadratic terms of the first derivatives. Expanding $n(r)$ in a Taylor series in powers of x,y,z we have

$$n_x(r) = a_1 x + a_2 y + a_3 z$$
$$+ \tfrac{1}{2}\left(a_{11}x^2 + a_{22}y^2 + a_{33}z^2 + a_{12}xy + a_{13}xz + a_{23}yz\right)$$
$$+ \cdots \cdots,$$
$$n_y(r) = 0, \qquad (6.4)$$
$$n_z(r) = \left[1 - n_x(r) - n_y(r)\right]^{\tfrac{1}{2}},$$

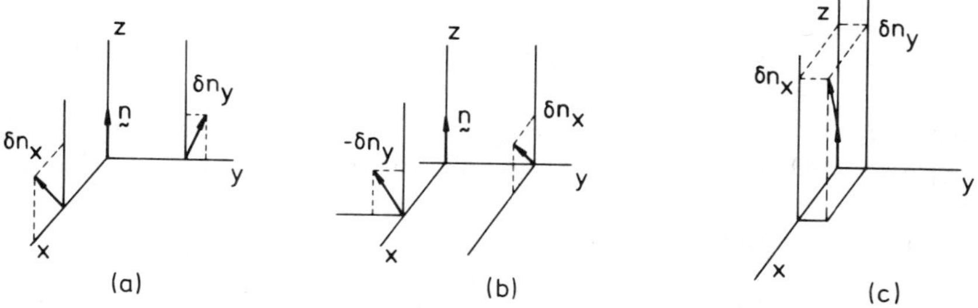

FIGURE 6.1 Splay (a), twist (b), and bend (c) deformations.

which leads to

$$a_1 = s_1, \qquad a_2 = t_2, \qquad a_3 = b_1,$$
$$a_{11} = ss, \qquad a_{22} = tt, \qquad a_{33} = bb,$$
$$a_{12} = st, \qquad a_{13} = sb, \qquad a_{23} = tb.$$

Since $n_x^2 + n_y^2 + n_z^2 = 1$ the first derivatives of n_z vanish, while the higher-order terms need not to be considered because they can be expressed in terms of the derivatives of n_x and n_y.

We now postulate the free-energy density of the nematic, relative to the free-energy density in the state of uniform alignment, to be both a quadratic function of the three curvature strains (6.2), and a linear function of the second-order terms (6.3):

$$F_{\text{dist}} = \sum_{i=1}^{3} k_i a_i + \tfrac{1}{2} \sum_{i,j=1}^{3} k_{ij} a_i a_j + \sum_{i,j=1}^{3} k_{ij}^{(2)} a_{ij}. \qquad (6.5)$$

A rotation about the z axis should make no change in the physical description of a uniaxial liquid crystal, and F_{dist} should be invariant for such rotations. This imposes restrictions on the elastic constants, and by considering a few of these rotations the number of independent constants in Eq. (6.5) can be reduced considerably (Frank, 1958). A further reduction follows by taking the equivalence of n and $-n$ into account. For example, of the second-order terms, only the splay-bend term sb is non-zero. Relieving the restriction to a planar configuration, we then find the following general expression for the free-energy density:

$$F_{\text{dist}} = k_2(t_1 + t_2) + k_{13}^{(2)}(d_1 + d_2),$$
$$+ \tfrac{1}{2} k_{11}(s_1 + s_2)^2 + \tfrac{1}{2} k_{22}(t_1 + t_2)^2 + \tfrac{1}{2} k_{33}(b_1^2 + b_2^2), \qquad (6.6)$$

where the second-order splay-bend terms have been denoted by d_1 and d_2. This expression can be further simplified as follows. The free energy being given by

$$\mathscr{F}_{\text{dist}} = \int F_{\text{dist}}\, dV, \qquad (6.7)$$

terms of the form ∇u, where $u(r)$ is an arbitrary vector field, can be transformed into surface integrals (see Gauss' theorem, appendix). Thus we can omit terms of this type in considerations involving the properties of a bulk nematic liquid crystal. Noting that

$$(d_1 + d_2) + (s_1 + s_2)^2 - (b_1^2 + b_2^2) = \nabla(n \cdot \nabla n), \qquad (6.8)$$

this applies to the $k_{13}^{(2)}$ term, provided that we rescale k_{11} and k_{33} as

$$K_1 = k_{11} - 2k_{13}^{(2)}, \qquad K_3 = k_{33} + 2k_{13}^{(2)}. \tag{6.9}$$

Eq. (6.6) differs from that in the literature in the absence of a term $(k_{22} + k_{24})(s_1 s_2 + t_1 t_2)$, which was missed because of our original assumption of a planar configuration. This is not important, however, as we can write

$$s_1 s_2 + t_1 t_2 = \frac{\partial n_x}{\partial x} \frac{\partial n_y}{\partial y} - \frac{\partial n_y}{\partial x} \frac{\partial n_x}{\partial y}$$

$$= \frac{\partial}{\partial x}\left(n_x \frac{\partial n_y}{\partial y}\right) - \frac{\partial}{\partial y}\left(n_x \frac{\partial n_y}{\partial x}\right). \tag{6.10}$$

Hence the term with $k_{22} + k_{24}$ also contributes to surface energies only. Writing $K_2 = k_{22}$, defining $t_0 = -k_2/K_2$, and adding a constant term $\frac{1}{2}K_2 t_0^2$ to Eq. (6.6), we thus arrive at:

$$F_{\text{dist}} = \frac{1}{2}\left[K_1(s_1 + s_2)^2 + K_2(t_1 + t_2 - t_0)^2 + K_3(b_1^2 + b_2^2) \right]. \tag{6.11}$$

Evidently t_0 is the twist of the equilibrium state. For the nematic phase, the free energy should be invariant to the choice of a left-handed or a right-handed coordinate system. This requirement leads to $k_2 = 0$ and thus to $t_0 = 0$. The chiral nematic phase is characterized by $t_0 \neq 0$.

Eq. (6.11) can be written in a vector notation, noting that

$$\nabla n = \partial n_x/\partial x + \partial n_y/\partial y = s_1 + s_2,$$

$$n \cdot \nabla \times n = \partial n_y/\partial x - \partial n_x/\partial y = -(t_1 + t_2), \tag{6.12}$$

$$(n \times \nabla \times n)^2 = (n \cdot \nabla n)^2 = (\partial n_x/\partial z)^2 + (\partial n_y/\partial z)^2 = b_1^2 + b_2^2.$$

Substitution in Eq. (6.11) leads, with $t_0 = 0$, to the fundamental equation of the continuum theory for nematics:

$$F_{\text{dist}} = \frac{1}{2}\left[K_1(\nabla \cdot n)^2 + K_2(n \cdot \nabla \times n)^2 + K_3(n \cdot \nabla n)^2 \right]. \tag{6.13}$$

The constants K_1, K_2 and K_3 are often referred to as Oseen-Frank constants. Our derivation differs from the usual one in the existence of the rescaling expression, Eq. (6.9), for K_1 and K_3 (Nehring and Saupe, 1971). Although this has no bearing on the final formula, Eq. (6.13), it is important if one wants to calculate K_1 and K_3 in terms of molecular models. In that case, the interaction energy of molecules with a different preferred orientation is considered, and a calculation of K_1 differs from a calculation of k_{11}. The experimentally observed stability of a uniform director pattern requires that K_1, K_2 and K_3 are all positive; this is not necessarily so for k_{11} and k_{33}.

To obtain the conditions for equilibrium in the bulk of a nematic, the total elastic energy associated with the distortion, Eq. (6.7), must be a minimum. In the presence of a magnetic field there is an extra term [see Eq. (3.8)]

$$F_{\text{magn}} = -\tfrac{1}{2}\mu_0^{-1}\Delta\chi\,(\boldsymbol{B}\cdot\boldsymbol{n})^2,\qquad(6.14)$$

where we have omitted the constant part of the magnetic energy that is independent of $\boldsymbol{n}$. In the electric case there will be an additional free energy density

$$F_{\text{diel}} = -\tfrac{1}{2}\varepsilon_0\Delta\varepsilon\,(\boldsymbol{E}\cdot\boldsymbol{n})^2.\qquad(6.15)$$

F_{dist} as given by Eq. (6.13) is a function of $\boldsymbol{n}(\boldsymbol{r})$ and of the gradients

$$g_{\alpha\beta} = \frac{\partial n_\beta}{\partial x_\alpha},\qquad \alpha,\beta = 1,2,3,\qquad(6.16)$$

where x_α stands for the three components x,y,z of $\boldsymbol{r}$. Formally, the condition of equilibrium can be expressed by saying that the total energy, obtained by substituting Eqs. (6.13), (6.14) and (6.15) in Eq. (6.7), is stationary with respect to variations of $\boldsymbol{n}(\boldsymbol{r})$ which keep $\boldsymbol{n}^2 = 1$. We recall that if one wants to make the value of an integral stationary with respect to variations of the integrand, the optimal function must satisfy the Euler-Lagrange equation (Margenau and Murphy, 1956, p. 200). We have to consider the functional derivative of F:

$$h_\alpha = -\frac{\partial F}{\partial n_\alpha} + \sum_\beta \frac{\partial}{\partial x_\beta}\frac{\partial F}{\partial g_{\beta\alpha}},\qquad \alpha = 1,2,3.\qquad(6.17)$$

The vector $\boldsymbol{h}$ that has been introduced has been called the molecular field. In equilibrium, the director must be at each point parallel to the molecular field. A review of the formal equilibrium theory of liquid crystals has been given by Ericksen (1976).

Various methods have been used to measure the elastic constants of nematic liquid crystals. We shall now discuss some of the most important of these.

(a) The Frederiks transition

The term Frederiks transition refers to the deformation of a layer with a uniform director pattern in an external field. Three configurations of practical importance are shown in Figure 6.2. As an example we shall discuss the situation of a uniform planar layer of thickness d (Figure 6.2a). The boundary conditions are assumed to be fixed: at $z = \pm\tfrac{1}{2}d$ the nematic is strongly anchored. A magnetic field is applied in the z direction. When

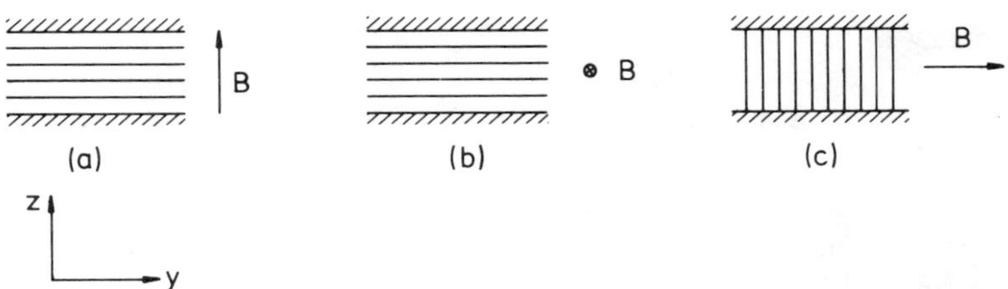

FIGURE 6.2 Splay, twist and bend modes for a Frederiks transition.

the field increases there is a gradual change in the director pattern once B exceeds a threshold value B_c (see Figure 6.3). For $B < B_c$ the equilibrium state is shown in Figure 6.3a (director everywhere in the y direction). In this case a small fluctuation of the director will be damped because the stabilizing elastic torque is larger than the destabilizing magnetic torque. For $B > B_c$ this situation has become an unstable equilibrium state; at the slightest fluctuation the system jumps to one of two stable states, (b) or (d) in Figure 6.3. In practice, the system chooses (b) in one domain, and (d) in another; the transitions between such domains (alignment-inversion walls) can be seen using a polarizing microscope. The tilt angle θ between the director in the middle of the layer and the xy plane varies from 0 to 90° if B increases from B_c without restriction (Figure 6.3a, b, c or a, d, e).

We shall now consider the effect described above in more detail. At any given field, the stable equilibrium state can be found by minimizing the total free energy with respect to variations in the director pattern. Let $\theta(z)$ be the tilt angle between the director and the xy plane. The director is then given by $n = [0, \cos\theta(z), \sin\theta(z)]$. For a unit surface of the layer the total

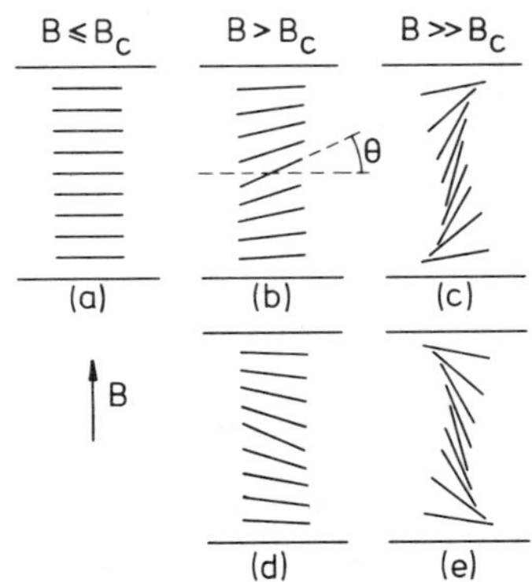

FIGURE 6.3 Deformation of the director pattern above the treshold in the case of the splay mode.

free energy is

$$\mathcal{F} = \int_0^{d/2} \left[(K_1 \cos^2\theta + K_3 \sin^2\theta)(\partial\theta/\partial z)^2 - \mu_0^{-1}\Delta\chi B^2 \sin^2\theta \right] dz. \quad (6.18)$$

The value of $\theta(z)$ will be such that $\mathcal{F}$ is a minimum with $\theta = 0$ for $z = \pm\frac{1}{2}d$. This condition leads to the Euler-Lagrange equation, that can be written directly as

$$(K_1 \cos^2\theta + K_3 \sin^2\theta)(\partial\theta/\partial z)^2 + \mu_0^{-1}\Delta\chi B^2 \sin^2\theta = C. \quad (6.19)$$

The constant C can be determined from the fact that $\theta(z)$ has a maximum value of θ_m for $z = 0$. Consequently at this position we have $\partial\theta/\partial z = 0$, and $C = \mu_0^{-1}\Delta\chi B^2 \sin^2\theta_m$. Thus one can write Eq. (6.19) as

$$B\, dz = \left[\frac{\mu_0}{\Delta\chi} \frac{K_1 + (K_3 - K_1)\sin^2\theta}{\sin^2\theta_m - \sin^2\theta} \right]^{\frac{1}{2}} d\theta. \quad (6.20)$$

We look for solutions other than $\theta(z) = 0$. First we restrict ourselves to small values of θ, and make the approximations $\sin^2\theta \approx \theta^2$ and $\sin^2\theta_m \approx \theta_m^2$. To calculate the threshold field we disregard the second term. In that case, Eq. (6.20) can be written as

$$B\, dz = (\mu_0 K_1/\Delta\chi)^{\frac{1}{2}}(\theta_m^2 - \theta^2)^{-\frac{1}{2}} d\theta. \quad (6.21)$$

This can readily be integrated from $\theta = \theta_m$ to $\theta = 0$ and $z = 0$ to $z = d/2$, giving the threshold field:

$$B_c d = \pi(\mu_0 K_1/\Delta\chi)^{\frac{1}{2}}. \quad (6.22)$$

For $B < B_c$ the minimum energy is obtained at $\theta(z) = 0$ for all values of z. The complete solution (Saupe, 1960) for arbitrary fields is

$$B/B_c = (2/\pi)(1 + \kappa \sin^2\theta_m)^{-\frac{1}{2}} \Pi(\alpha^2, k), \quad (6.23)$$

where

$$\kappa = (K_3 - K_1)/K_1,$$

$$\alpha^2 = \kappa \sin^2\theta_m/(1 + \kappa \sin^2\theta_m),$$

$$k^2 = \sin^2\theta_m(1 + \kappa)/(1 + \kappa \sin^2\theta_m),$$

and $\Pi(\alpha^2, k)$ is the complete elliptic integral of the third kind. As we see from Eq. (6.20), for $B > B_c$ the deformation depends on the ratio K_3/K_1.

If we start with a homeotropic layer and apply the field in the xy plane (Figure 6.2c), the above results remain valid provided we replace K_1 by K_3

and *vice-versa*, or alternatively θ by $\frac{1}{2}\pi-\theta$. If we start with a uniform planar layer and apply the field in the x direction (Figure 6.2b), the situation is slightly different. In that case a pure twist occurs; for all values of the field the deformation depends of K_2 only. The deformation is given (Saupe, 1960) by

$$B/B_c=(2/\pi)K(\sin^2\phi),\tag{6.24}$$

where ϕ is the twist angle, and $K(\sin^2\phi)$ the complete elliptic integral of the first kind. In conclusion, we can write Eq. (6.22), generalized for the three geometries of Figure 6.2 (indicated as the splay, twist and bend modes) as

$$\mu_0^{-1}\Delta\chi B_c^2 d^2=K_i\pi^2,\qquad i=1,2,3.\tag{6.25}$$

Another situation that is of practical interest is the twisted planar layer, which can be obtained by rotating the upper boundary of a uniform planar layer through an angle ϕ (see Figure 2.2c for $\phi=\pi/2$). Application of a field in the z direction leads again to a distortion of the director pattern with a threshold field given (Leslie, 1970) by

$$\mu_0^{-1}\Delta\chi B_c^2 d^2=K_1\pi^2+(K_3-2K_2)\phi^2.\tag{6.26}$$

For fields $B\gg B_c$ the optical properties of such a sample change drastically, and this makes the effect suitable for display applications. In that case, of course, an electric field will be used instead of a magnetic field.

From the above discussion, it is clear that the elastic constants of a particular nematogenic compound can be obtained by measuring the threshold values for the three appropriate geometries. In principle any anisotropic property, such as the dielectric permittivity or the electric or thermal conductivity can be used to probe the average state of alignment. As the probing field can usually be applied only normal to the layer, these methods work only for the splay and bend modes, and not in the twist geometry.

The distortion of the director pattern is most accurately detected optically. Considering again the splay geometry, we note that a light wave polarized along the x axis travels with the ordinary refractive index n_o, irrespective of the distortion. If polarized along the y axis the refractive index is n_e for $B<B_c$, but otherwise one has locally

$$n(z)=n_e n_o\left(n_e^2\sin^2\theta+n_o^2\cos^2\theta\right)^{-\frac{1}{2}}.\tag{6.27}$$

A light beam polarized in an arbitrary direction will be split into an ordinary and an extraordinary ray, which can be made to interfere at an analyzer. The difference in optical path length is

$$\delta = (2/\lambda) \int_0^{d/2} \left[n(z) - n_o \right]. \tag{6.28}$$

Hence the transmitted light will show a series of minima and maxima, depending on whether δ equals an integer or an integer plus one half. This is illustrated in Figure 6.4. A similar discussion applies to the bend mode. The solutions of Eq. (6.28) are discussed by Saupe (1960) and Gruler *et al.* (1972). In the twist geometry, there is no phase difference for normally incident light. The sample can be divided into two halves that give equal contributions to δ, but of opposite sign. In order to detect the transition, one has to use light incident at an angle to the normal to the layer. The most elegant method is to monitor the conoscopic figure. The characteristic conoscopic figure of a uniform planar layer, a set of hyperbolae with one axis parallel to the director, starts to rotate when $B > B_c$ (Cladis, 1972).

In practice there are various complications when Frederiks transitions are being studied experimentally. We shall briefly mention some of them.

i) From Eq. (6.25), we see that B_c is inversely proportional to d, the thickness of the nematic layer, as originally observed by Frederiks. At the boundaries, strong anchoring is assumed. If this is not the case the deformation will be as if the thickness were $d + b$, instead of d, with a correspondingly lower value of B_c. The extrapolation length b is a measure of the surface energy and can be determined from measurements at

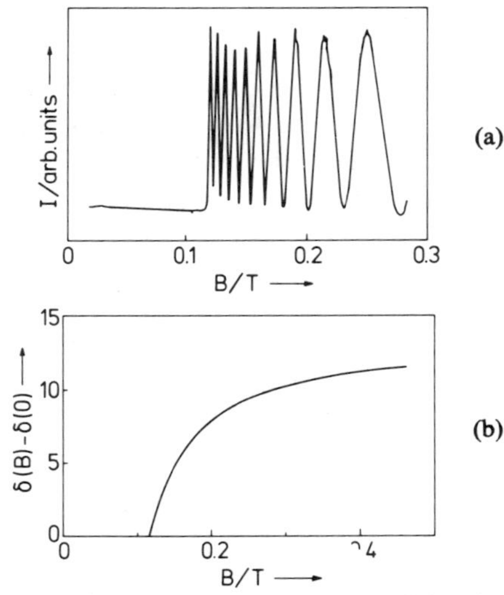

FIGURE 6.4 (a) Maxima and minima in the transmitted light above the treshold of a Frederiks transition (splay mode); (b) phase difference for the same configuration.

various thicknesses. Experimentally, different surface treatments indeed give different values for B_c (de Jeu *et al.* 1976).

ii) The theory discussed so far applies to the situation $\boldsymbol{B} \perp \boldsymbol{n}$. If this condition is not fulfilled there is no longer any real threshold. The theory for the case of oblique fields has been reviewed by Deuling (1978). For even small angles between $\boldsymbol{B}$ and the normal to $\boldsymbol{n}$ (say $\sim 1°$) there is already a rounding off in the curve of δ *versus* B (see Figure 6.4b) near B_c. This makes the extrapolation to obtain B_c difficult, and leads to values of B_c that are too small.

iii) When an electric field is applied, the situation is in many respects as discussed above (Gruler *et al.* 1972). However, complications may arise because of three effects. Firstly, due to the anisotropic conductivity, space charge may be generated in a distorted director pattern, which has some influence on the transition (Gruler and Cheung, 1975). Secondly, even if the nematic were non-conducting, a distortion of the director pattern can induce an electric polarization. This effect has been called flexo-electricity, the curvature-dependent analogue of piezo-electricity in solids (see, for example, Deuling, 1978). Finally, above the threshold, the dielectric field is not constant throughout the layer.

In principle, it would be possible to obtain both K_1 and K_3 from a single experiment; for example, K_1 may be obtained from the threshold value, and K_3/K_1 from a fit of the experimental and theoretical curve of δ *versus* B. This requires accurate values of the refractive indices, as the fit is very sensitive to the value of Δn. The above mentioned problems, however, make the results from such a fitting procedure somewhat doubtful, and usually a check will be required with the complementary mode. Anyhow, all these effects indicate that a high value of a specific elastic constant (corresponding to a high threshold field) is in general more trustworthy than a lower one.

(b) Light scattering

The molecules in a liquid crystal tend to be parallel to the director, around which they fluctuate as measured by the order parameter S. However, in a region of space the director $\boldsymbol{n}$ may differ from its equilibrium orientation $\boldsymbol{n}_0$ by an amount

$$\delta \boldsymbol{n} = \boldsymbol{n} - \boldsymbol{n}_0. \tag{6.29}$$

The small fluctuation $\delta \boldsymbol{n}$ must be perpendicular to $\boldsymbol{n}_0$ since necessarily $\boldsymbol{n}^2 = 1$. These fluctuations can be described in terms of the curvature elastic theory; hence they depend on the elastic constants. As it costs no energy to rigidly rotate the director, the energy required to create an orientational fluctuation of n is small for long wavelength λ of the fluctuation. Since the polarizability of the medium is anisotropic, these fluctuations give rise to fluctuations in the optical dielectric permittivity, and thus to light scattering. Introducing the wave vector q with $q = 2\pi\bar{n}/\lambda$, where $\bar{n}$ is the average refractive index, the scattering will be strong for small values of q, causing the turbid appearance of the nematic. Above T_{NI} these long wavelength modes no longer exist; the only fluctuations in the permittivity then are due to density fluctuations. As these are energetically much less favourable, a clear liquid results.

To discuss the fluctuations of n in more detail we consider a nematic sample with n_0 parallel to the z axis. Hence the fluctuations at any point r can be described by $n_x(r)$ and $n_y(r)$. The total curvature elastic energy is

$$\mathcal{F}_{dist} = \tfrac{1}{2}\int\left\{K_1\left(\frac{\partial n_x}{\partial x} + \frac{\partial n_y}{\partial y}\right)^2 + K_2\left(\frac{\partial n_x}{\partial y} - \frac{\partial n_y}{\partial x}\right)^2 \right.$$
$$\left. + K_3\left[\left(\frac{\partial n_x}{\partial z}\right)^2 + \left(\frac{\partial n_y}{\partial z}\right)^2\right]\right\}dV. \tag{6.30}$$

As in the theory of thermal motion of atoms in solids, it is convenient to analyse the fluctuating quantities $n_x(r)$ and $n_y(r)$ in terms of the Fourier components:

$$n_x(r) = \sum_q n_x(q)\exp(iq\cdot r), \tag{6.31}$$

and similarly for $n_y(r)$. The inverse transformation of Eq. (6.31) reads

$$n_x(q) = V^{-1}\int n_x(r)\exp(-iq\cdot r)\,dV, \tag{6.32}$$

where V is the sample volume. Hence one finds

$$(\partial n_x/\partial x)^2 = \sum_q |n_x(q)|^2 q_x^2, \; etc. \tag{6.33}$$

It is convenient to rotate for each q the coordinate system x,y,z around the z axis, such that the new x' axis coincides with the unit vector e_1, which is perpendicular to the z axis in the qz plane (see Figure 6.5). The y' axis then coincides with e_2. In this new system we have $q_y = 0$, and the components of $n(q)$ along e_α are $n_\alpha(q)$ ($\alpha = 1, 2$). Using Eq. (6.31) the free energy then takes the simple form

$$\mathcal{F}_{dist} = \tfrac{1}{2}V\sum_q\left[|n_1(q)|^2(K_1 q_\perp^2 + K_3 q_\parallel^2) + |n_2(q)|^2(K_2 q_\perp^2 + K_3 q_\parallel^2)\right],$$
$$\tag{6.34}$$

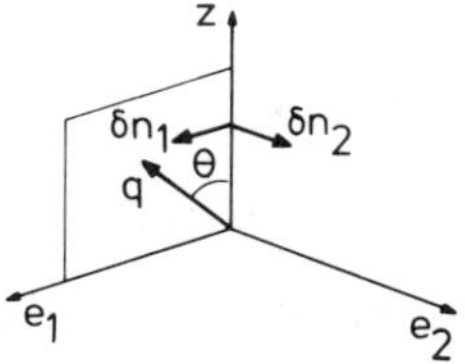

FIGURE 6.5   Definition of the eigen-modes for long wavelength fluctuations.

where $q_\parallel = q_z = q\cos\theta$ and $q_\perp = q\sin\theta$ (see Figure 6.5). To calculate $\langle |n_\alpha(\boldsymbol{q})|^2 \rangle$, the thermal average of $|n_\alpha(\boldsymbol{q})|^2$, one can use the equipartition theorem. For a classical system in thermal equilibrium, the average energy, per degree of freedom, is equal to $\frac{1}{2}k_B T$. Hence we arrive at

$$\langle |n_\alpha(\boldsymbol{q})|^2 \rangle = (k_B T/V)\left(K_\alpha q_\perp^2 + K_3 q_\parallel^2\right)^{-1}, \qquad \alpha = 1,2. \qquad (6.35)$$

This is the central equation of fluctuation theory for nematics (de Gennes, 1968). Thanks to the use of the equipartition theorem, an expression for the variance of the thermal fluctuations of $\boldsymbol{n}$ can be given, that is independent of the dynamic processes that in fact govern the fluctuations of $\boldsymbol{n}$.

We now come to the effect of the fluctuations in $\boldsymbol{n}$ on the optical dielectric tensor. Let us consider the geometry of Figure 6.6 as an example. The electric field of the incident light beam can be described by

$$E_{\text{in}}(\boldsymbol{r}) = iE_0\exp(i\boldsymbol{k}_0\cdot\boldsymbol{r}), \qquad (6.36)$$

where E_0 is the amplitude, and $\boldsymbol{i}$ a unit vector perpendicular to the ingoing wave vector $\boldsymbol{k}_0$, defining the direction of polarization. Due to this electric field the nematic sample acquires a polarization $\boldsymbol{P}(\boldsymbol{r})$, related to $E_{\text{in}}(\boldsymbol{r})$ by Eq. (5.1). We observe the sample from some distant point $\boldsymbol{r}' = \boldsymbol{r} + \boldsymbol{R}$, where $\boldsymbol{R}$ is large compared with the size of the sample. The interest then is in the total scattering field $E_{\text{out}}(\boldsymbol{r}')$ in an outgoing wave of wave vector $\boldsymbol{k}'\|\boldsymbol{R}$, at a polarization given by the unit vector $\boldsymbol{f}$. Then the relevant element of the

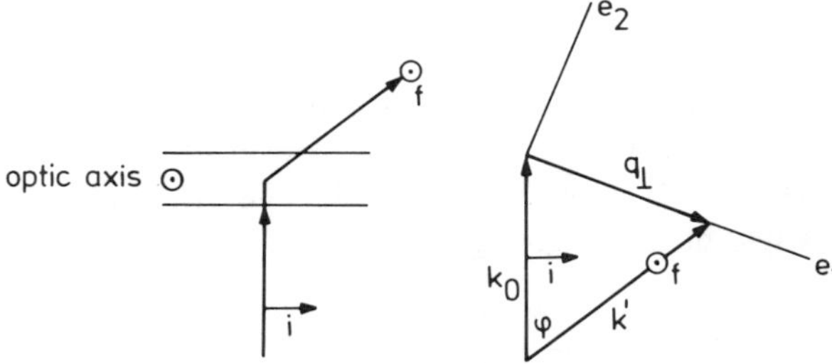

FIGURE 6.6   Scattering configuration with the incident beam polarized perpendicular to the director.

optical dielectric tensor is

$$\varepsilon_{if} = f \cdot \varepsilon \cdot i$$

$$= \bar{\varepsilon} + \Delta\varepsilon\left(n_i n_f - \tfrac{1}{3}\right), \tag{6.37}$$

where $n_i = n \cdot i$ and $n_f = n \cdot f$ are the components of n along the two polarization directions. Using Eq. (6.29) we find for the fluctuating part of ε_{if}:

$$\delta\varepsilon_{if} = \Delta\varepsilon\left[\, i_{\parallel}(f \cdot \delta n) + f_{\parallel}(i \cdot \delta n)\,\right], \tag{6.38}$$

where $i_{\parallel} = n_0 \cdot i$, etc. Analyzing δn in the eigenmodes n_1, n_2 defined above:

$$\delta n(q) = e_1 n_1(q) + e_2 n_2(q), \tag{6.39}$$

we then find

$$\delta\varepsilon_{if} = \Delta\varepsilon \sum_{\alpha = 1,2} n_\alpha(q)(i_\alpha f_{\parallel} + i_{\parallel} f_\alpha), \qquad \alpha = 1,2, \tag{6.40}$$

where $i_\alpha = i \cdot e_\alpha$ is the component of i along e_α.

At this stage we see that the orientation fluctuations couple *via* the optical dielectric tensor to the incident electric field, giving a fluctuating polarization $P(r)$. From classical electromagnetic theory the field radiated in the direction $R = r' - r$ by an electric dipole $P(r)$, oscillating at angular frequency ω, is known to be

$$E(r + R) = \left[\omega^2 / \left(4\pi\varepsilon_0 c^2 R\right)\right] P_\nu(r)\exp(ikR), \tag{6.41}$$

where $P_\nu(r)$ is the component of $P(r)$ normal to R. For simplicity we assume that the optical anisotropy is small and take $k = \bar{n}\omega/c$, where $\bar{n}$ is the average refractive index, and c the velocity of light in a vacuum. The outgoing field $E_{\text{out}}(r')$ is obtained by integrating all contributions over the volume of the sample. For large R the factor $1/R$ can be taken out of the integral. Writing $kR = k' \cdot R = k' \cdot (r' - r)$, where k' is the wave vector of the outgoing beam, we find using Eqs. (5.1) and (6.36)

$$f \cdot E_{\text{out}}(r') = (\omega^2 / 4\pi c^2)(E_0 / R)\exp(ik' \cdot r') \int \delta\varepsilon_{if}(r)\exp(-iq \cdot r)\, dV,$$

$$= A(E_0 / R)\exp(ik' \cdot r'), \tag{6.42}$$

where $q = k' - k_0$, and A is the scattering amplitude. The integral is, apart from a factor V, precisely the Fourier transform of $\delta\varepsilon_{if}(r)$ which can be written as $\delta\varepsilon_{if}(q)$, and is given by Eq. (6.40). The differential cross-section per unit solid angle of the outgoing beam around the direction k' is

$$\frac{d\sigma}{d\Omega} = \langle|A|^2\rangle = \left(\frac{\Delta\varepsilon\,\omega^2 V}{4\pi c^2}\right)^2 \sum_{\alpha = 1,2} \langle|n_\alpha(q)|^2\rangle(i_\alpha f_{\parallel} + i_{\parallel} f_\alpha)^2, \tag{6.43}$$

where the brackets indicate the thermal average given by Eq. (6.35).

From the polarization factor in Eq. (6.43), we see that the intensity of the scattered light has a maximum for crossed polars, in contrast to the situation in isotropic liquids. Let us consider as an example the geometry sketched in Figure 6.6. The incident and the scattered beam are both normal to the z axis. The incident beam is linearly polarized in the scattering plane, while the scattered beam is polarized along z, the optical axis of the nematic. The angle of scattering being denoted by ϕ, we have

$$\tfrac{1}{2}q_\perp \approx k\sin(\phi/2), \qquad q_\parallel = 0,$$

$$i_1 = \cos(\phi/2), \qquad i_2 = \sin(\phi/2),$$

$$i_\parallel = f_1 = f_2 = 0, \qquad f_\parallel = 1.$$

Substitution in Eq. (6.43) gives:

$$\frac{d\sigma}{d\Omega} = \frac{\Delta\varepsilon^2\omega^4}{16\pi^2 c^4}k_B T\left[\frac{\cos^2(\phi/2)}{K_1 q_\perp^2} + \frac{\sin^2(\phi/2)}{K_2 q_\perp^2}\right]$$

$$\sim \cot^2(\phi/2) + K_1/K_2. \tag{6.44}$$

Experimentally, this functional relation is indeed approximately obeyed. Hence by extrapolating the straight line one can in principle obtain a value for the ratio K_1/K_2. In practice corrections have to be made for the birefringence of the nematic sample. To obtain a straight line then Eq. (6.44) has to be modified somewhat (see, for example, van Eck and Zijlstra, 1978). The result is illustrated in Figure 6.7. In a similar way K_3/K_2 can be obtained by studying the dependence of the cross-section on the angle between q and the optical axis.

With a strong magnetic field the fluctuations $\langle|n_\alpha(q)|^2\rangle$ are decreased; they become inversely proportional to $K_\alpha q_\perp^2 + K_3 q_\parallel^2 + \mu_0^{-1}\Delta\chi B^2$ (compare

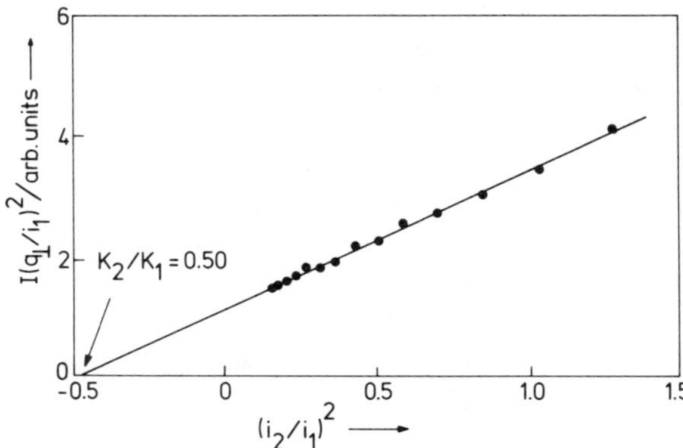

FIGURE 6.7 Determination of K_2/K_1 from the intensity of scattered light as a function of the scattering angle (p-methoxy-p'-butylazoxybenzene at 47.5°C, courtesy D. C. van Eck).

Eq. 6.35). If $\Delta\chi$ is known, it then becomes possible to measure the magnitudes of the three elastic constants from relative intensity measurements at various wave vectors.

(c) Miscellaneous methods

In this paragraph we shall briefly discuss some other methods. First we mention the *cholesteric-nematic transition*. If a magnetic field of flux-density $\boldsymbol{B}$ is applied perpendicular to the helix axis of a chiral nematic, the pitch is found to increase when $\Delta\chi$ is positive. There is a critical field given by

$$\mu_0^{-1}\Delta\chi B_c^2 = \tfrac{1}{4}\pi^2 t_0 K_2, \tag{6.45}$$

above which the liquid crystal has nematic ordering. The pitch diverges logarithmically as B_c is approached. The measurement of B_c gives information on K_2, provided the pitch and $\Delta\chi$ are known. In practice, Eq. (6.45) has been used for long-pitch chiral nematics, obtained by doping a nematic with enantiomorphic molecules (see, for example, Durand *et al.* 1969). The drawback of this method is that K_2 of the *pure* nematic can not be obtained.

The nematic state is named after the apparent threads seen within the fluid under a microscope. These are line singularities or *disclinations*. The director rotates by a multiple of π on a circuit taken around one of the lines. The actual configuration around disclination lines can be calculated with the aid of the elastic continuum theory. For some types of disclination, this configuration is sensitive to whether $K_1/K_3 \gtrless 1$ (Frank, 1958). This provides a means of obtaining information on the relative magnitude of K_1 and K_3 which hardly seems to have been explored. Defects and textures will not be discussed here any further. An excellent introduction to this subject can be found in Chapter 4 of de Gennes' book.

Finally *alignment-inversion walls* can be used to study the elastic constants. If a field is applied normal to a uniform nematic layer, the distortion above the threshold can be achieved in two equivalent ways, corresponding to an angle between the field and the director of $+\theta$ or $-\theta$ (see Figure 6.3). Two such adjacent domains are separated by a wall, in which the director rotates continuously from $+\theta$ to $-\theta$. If the field is suddenly raised to a value above B_c, closed domains are often generated. The equilibrium shape of a wall surrounding a closed domain is an ellipse. In the splay and the bend geometry, the axial ratio of this ellipse equals $(K_3/K_2)^{\frac{1}{2}}$ and $(K_1/K_2)^{\frac{1}{2}}$, respectively (Brochard, 1972). A closed wall will spontaneously collapse, but as long as the size of the ring is larger than the wall thickness, its shape remains an ellipse of fixed ellipticity. In this way the ratio of two elastic constants can be determined (Leger, 1972).

We now come to the-discussion of some typical results for the elastic constants. The most extensive studies have been made using the technique of the Frederiks transition. However, most of the older results are not very trustworthy, because the crucial role of the anchoring conditions was not fully appreciated. This is illustrated in Figure 6.8, where results for $K_i/\Delta\chi$ from various sources are shown for PAA and MBBA. As we see, different surface treatments give different thresholds. One of the classical methods by which to obtain homeotropic alignment is to wash the substrate with chromic acid. For PAA, the results for $K_3/\Delta\chi$ from three different sources, all using this alignment method, practically coincide. However, after a different surface treatment higher values were obtained. In the absence of independent information on the anchoring conditions, we can only say that the highest values for K_1 and K_3 are the most trustworthy. It seems that it is much easier to obtain strong anchoring in the twist mode. In this geometry, the anchoring conditions can easily be checked by considering the agreement between theory and experiment for the rotation of the conoscopic figure at $B > B_c$ (Leenhouts et al. 1976). In the splay and bend modes, such a check is also possible in principle, but it depends on the accuracy with which Δn is known, thus introducing a further uncertainty. The elastic constants are of the order of 10^{-11} newton. As $K \sim U/a$, where U is an interaction energy and a is a molecular dimension, this is about what one would expect for $U \approx 10 \text{ kJ mol}^{-1}$ (≈ 0.1 eV) and $a \approx 1.5$ nm.

In Table 6.1 we give a set of elastic constants for some compounds. For K_1 and K_3 of MBBA and PAA we have taken the value for $K_i/\Delta\chi$ from the

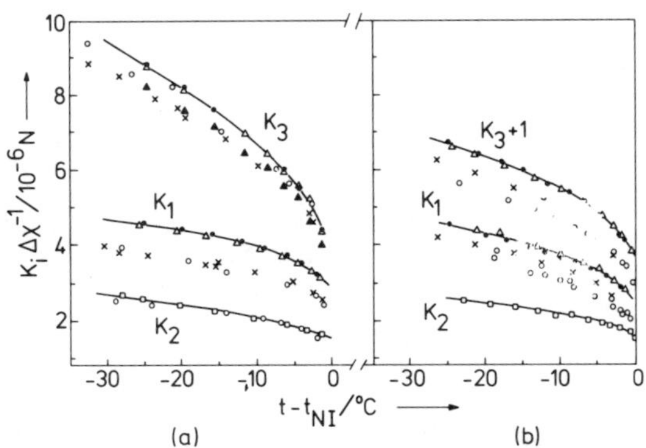

FIGURE 6.8   Elastic constants of PAA (a) and MBBA (b). Different symbols correspond to different sources and surface treatments as indicated by de Jeu et al. (1976).

TABLE 6.1
Elastic constants for some nematic liquid crystals (10^{-12}N).

Compound	$t_{NI}(°C)$	$t_{NI}-t(°C)$	K_1	K_2	K_3
PAA	135.5	13.5	6.9[a]	3.8[b]	11.9[a]
MBBA	47	23	7.1[a]	4.0[c]	9.2[a]
		18	6.4[a]	3.6[c]	8.2[a]
DIBAB	32	10	4.8[d]	3.0[d]	4.7[d]

[a]de Jeu *et al.* (1976); [b]Madhusudana *et al.* (1975);
[c]Leenhouts *et al.* (1976); [d]de Jeu and Claassen (1977)

highest thresholds given in Figure 6.8. The magnetic susceptibilities and densities used to calculate the K_i have been taken from de Jeu *et al.* (1976). In the case of MBBA a comparison with results from other methods is possible. From light scattering experiments Martinand and Durand (1972) obtained $K_3 = (7.2 \pm 1) \times 10^{-12}$ N. The difference compared with the value in Table 6.1 is probably due to a slightly smaller value of $T_{NI} - T$ in their experiment. The value $K_3/K_1 = 1.3$ from Table 6.1 is in good agreement with the value 1.38 obtained from the axial ratio of inversion walls (Leger, 1972).

The limited amount of systematic experimental results available for the elastic constants suggests that geometrical factors as the length L and the width D of the molecules have a great influence on these properties. This is illustrated in Figure 6.9, which shows that for some compounds that differ in their end or bridging group, but otherwise have approximately the same length and width, K_1 is very similar. The ratio K_3/K_1 does not differ much in these cases either. This trend has been substantiated by Schadt and

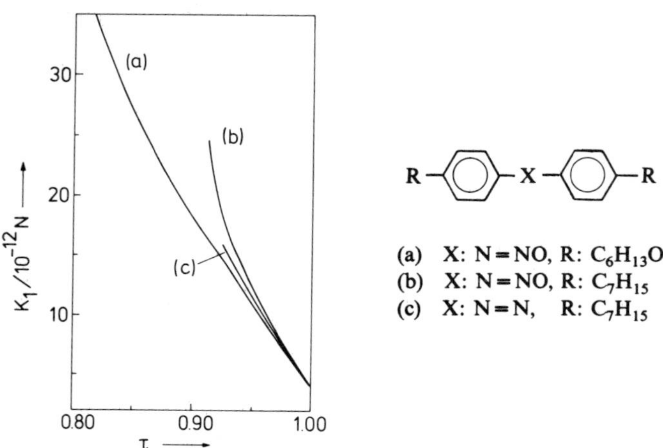

FIGURE 6.9 Elastic constants K_1 for some different compounds with the same length-width ratio.

Müller (1979), who found little variation in the values of both K_1 and K_3 for a large number of compounds with variable central parts but the same end groups. When summarizing the experimental trend it is useful to distinguish between relatively rigid molecules, and molecules containing flexible groups.

(i) Rigid molecules. The first three examples of Table 6.2 show that if the length of the molecules is increased by adding relatively rigid groups, K_3/K_1 increases. The reverse effect is observed when the width of a molecule is increased. Adding an *o*-hydroxy group to MBBA leads to an increase of K_1, while K_3 remains approximately constant (Leenhouts *et al.* 1979b). These observations suggest

$$K_3/K_1 \sim L/D. \qquad (6.46)$$

As we shall see shortly, there are theoretical reasons to expect such a relationship.

(ii) Molecules with flexible groups. When an alkyl or alkoxy chain is added to the aromatic core of a molecule, K_3/K_1 is found to decrease

TABLE 6.2
Ratio K_3/K_1 for some compounds with a different length/width ratio ($T/T_{NI} = 0.96$)[a]

No.	Compound	K_3/K_1
1	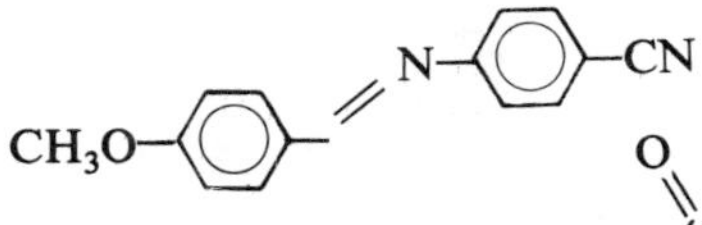 CH$_3$O—⟨ring⟩—=N—⟨ring⟩—CN	1.9
2	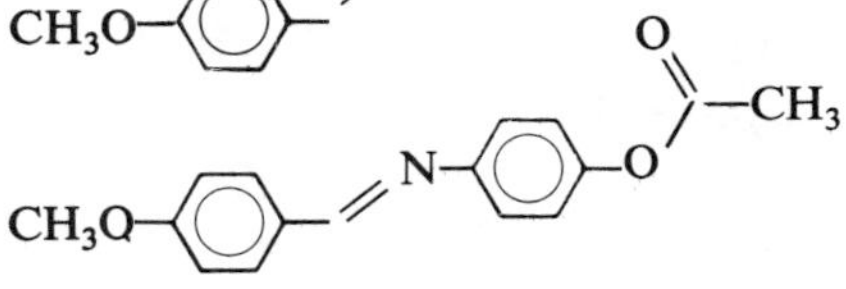 CH$_3$O—⟨ring⟩—=N—⟨ring⟩—O—C(=O)—CH$_3$	2.1
3	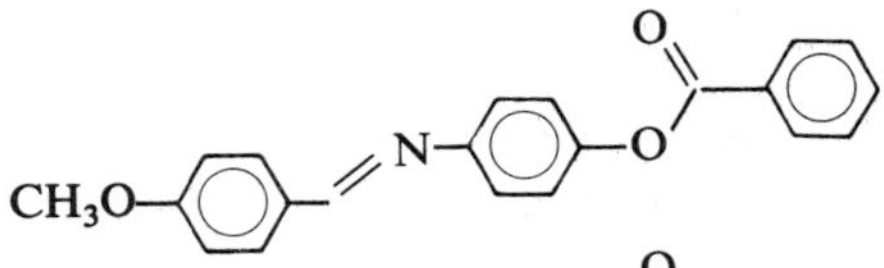 CH$_3$O—⟨ring⟩—=N—⟨ring⟩—O—C(=O)—⟨ring⟩	2.4
4	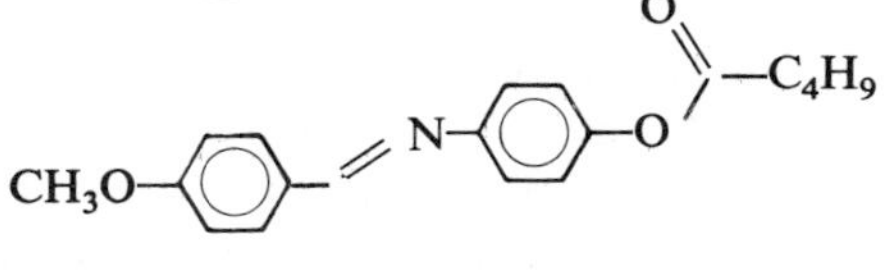 CH$_3$O—⟨ring⟩—=N—⟨ring⟩—O—C(=O)—C$_4$H$_9$	1.3

[a]Leenhouts *et al.* (1979a)

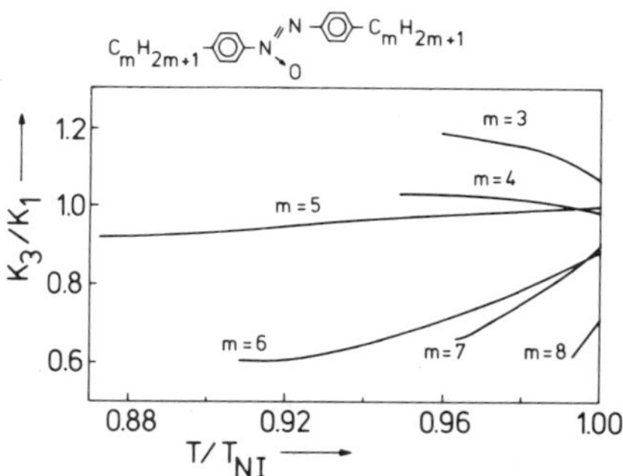

FIGURE 6.10   Ratio K_3/K_1 for the series of p, p'-dialkylazoxybenzenes (de Jeu and Claassen, 1977).

(compare no. 4 of Table 6.2 with nos. 2 and 3). This effect is illustrated for a homologous series in Figure 6.10. The decrease of K_3/K_1 with increasing chain length is due to the fact that K_1 increases, while K_3 does not change much.

For K_2 there is even less information available than for K_1 and K_3. Up to now, K_2 has always been found to be the smallest of the three constants. The ratio K_2/K_1 varies for various compounds from 0.4 to about 0.8, K_2 being remarkably constant.

In theoretical studies of curvature elasticity, often the approximation $K_1 = K_2 = K_3$ is made. From the results we see that this one-constant approximation bears little relation with physical reality. On the other hand, compounds can be found for which $K_1 \approx K_3$. Then a two-constant approximation can be used, which may result in a considerable simplification of the equations (see, for example, Eqs. 6.18–6.20). Table 6.1 includes an example of such a compound, p,p'-dibutylazoxybenzene (DIBAB). For this compound the ratio K_3/K_1 has also been determined by light scattering. The value of 1.25 ± 0.2, although somewhat higher than that measured *via* the Frederiks transition, agrees reasonably well with that from Table 6.1.

Before attempting to interpret the above results in terms of molecular models, we shall discuss the behaviour of the elastic constants as a function of the order parameter. In the Landau theory of the nematic phase, the free energy $F(p,T,Q_{\alpha\beta})$ is assumed to be an analytical function of the order parameter tensor **Q**, Eq. (6.1). To the extent that $Q_{\alpha\beta}$ is small, we can expand the free energy in the various orders of $Q_{\alpha\beta}$. When curvature is involved in the director pattern, the tensor order parameter will vary from point to point, and terms in the gradients of $Q_{\alpha\beta}$ will also occur in the expansion. The free energy up to the second order in $Q_{\alpha\beta}$ can then be written (de Gennes, 1971) as:

$$F = F_i + \frac{1}{2} A \sum_{\alpha,\beta} Q_{\alpha\beta} Q_{\alpha\beta}$$

$$+ \sum_{\alpha,\beta,\gamma} \left[\frac{1}{2} L_1 \left(\frac{\partial Q_{\beta\gamma}}{\partial x_\alpha} \right) \left(\frac{\partial Q_{\beta\gamma}}{\partial x_\alpha} \right) + \frac{1}{2} L_2 \left(\frac{\partial Q_{\alpha\gamma}}{\partial x_\alpha} \right) \left(\frac{\partial Q_{\beta\gamma}}{\partial x_\beta} \right) \right], \quad (6.47)$$

where x_α ($\alpha = 1,2,3$) stands for x,y,z and F_i is the free energy density of the isotropic phase. Only two elastic constants occur, denoted by L_1 and L_2. In order to relate L_1 and L_2 to the Oseen-Frank constants, we can substitute Eq. (6.1) into Eq. (6.47). Assuming that the variations of $Q_{\alpha\beta}$ are due to variations of $\boldsymbol{n}$ only, and that S is spatially constant, the part of the free energy due to the distortion reduces, after some manipulations (Stephen and Straley, 1974), to

$$F_{\text{dist}} = S^2 \left[\left(L_1 + \tfrac{1}{2} L_2 \right)(\nabla \cdot \boldsymbol{n})^2 + L_1 (\boldsymbol{n} \cdot \nabla \times \boldsymbol{n})^2 + \left(L_1 + \tfrac{1}{2} L_2 \right)(\boldsymbol{n} \cdot \nabla \boldsymbol{n})^2 \right].$$

$$(6.48)$$

Comparison with Eq. (6.13) gives

$$K_1 = K_3 = 2S^2 \left(L_1 + \tfrac{1}{2} L_2 \right),$$
$$K_2 = 2S^2 L_1. \qquad (6.49)$$

Hence, any theory that contains no terms of higher order than S^2 will predict $K_1 = K_3$. Moreover, the elastic constants then must vary with temperature as S^2. This temperature variation has often been assumed to be *generally* valid, which, as we see from the above Landau-expansion, is not true. A temperature variation $K_i \sim S^2$ can only be expected if the two-constant approximation is valid. From the experimental results, especially Figure 6.10, we see that these predictions are in excellent agreement with experiment. If $K_1 \approx K_3$ the ratio K_3/K_1 is almost independent of temperature. With increasing temperature S decreases, and the errors due to disregarding higher order terms become smaller. Indeed, the temperature dependence of K_3/K_1 in the case $K_3/K_1 \gtrsim 1$ is such that with increasing temperature the values of K_1 and K_3 approach each other.

A molecular approach to the elastic constants can be made by considering the interaction energy between the molecules in two volume elements dV and dV' in a nematic liquid crystal, the preferred orientation in these volume elements being different. Following essentially Nehring and Saupe (1971), this interaction energy can be taken equal to

$$f \, dV \, dV'. \qquad (6.50)$$

At fixed temperature and pressure, the function f is assumed to depend only on the relative position of the volume elements, and on the directions of n and $n' = n + \delta n$, which vectors give the preferred molecular orientation at dV and dV', respectively. We shall take dV at the origin of a coordinate system x, y, z with n along the z axis. The direction of $n' = n + \delta n$ is taken in the xz plane at a small angle α with the z axis (see Figure 6.11). This particular choice implies $\delta n \approx \delta n_x$, $\delta n_z \approx 0$ and $\delta n_y = 0$. From Eqs. (6.2) and (6.3) we see that this choice gives no loss of generality, as with δn_x one can construct all types of deformation. As n and n' are unit vectors we have for small α

$$\delta n_x = \sin\alpha \approx \alpha. \tag{6.51}$$

Now we shall assume that the interactions decrease rapidly with distance, and can be ignored for distances greater than a maximum distance R. For all volume elements dV' within this distance from the origin, we can expand f into a power series in α:

$$f = f_0 + \left(\frac{\partial f}{\partial \alpha}\right)_{\alpha=0} \alpha + \frac{1}{2}\left(\frac{\partial^2 f}{\partial \alpha^2}\right)_{\alpha=0} \alpha^2 + \cdots \tag{6.52}$$

On the other hand α can be expanded in a power series of the relative coordinates x, y, z of the element dV':

$$\alpha = \sum_{i=1}^{3}\left(\frac{\partial \alpha}{\partial x_i}\right)_{x_i=0} x_i + \frac{1}{2}\sum_{i,j=1}^{3}\left(\frac{\partial^2 \alpha}{\partial x_i \partial x_j}\right)_{x_i, x_j=0} x_i x_j + \cdots, \tag{6.53}$$

where x_i ($i = 1, 2, 3$) has been written for x, y, z. Substitution in Eq. (6.52) gives

$$\begin{aligned}
f = f_0 &+ \sum_{i=1}^{3}\left(\frac{\partial f}{\partial \alpha}\right)_{\alpha=0}\left(\frac{\partial \alpha}{\partial x_i}\right)_{x_i=0} x_i \\
&+ \frac{1}{2}\sum_{i,j=1}^{3}\left[\left(\frac{\partial^2 f}{\partial \alpha^2}\right)_{\alpha=0}\left(\frac{\partial \alpha}{\partial x_i}\right)_{x_i=0}\left(\frac{\partial \alpha}{\partial x_j}\right)_{x_j=0} \right. \\
&\left. + \left(\frac{\partial f}{\partial \alpha}\right)_{\alpha=0}\left(\frac{\partial^2 \alpha}{\partial x_i \partial x_j}\right)_{x_i, x_j=0}\right] x_i x_j + \cdots
\end{aligned} \tag{6.54}$$

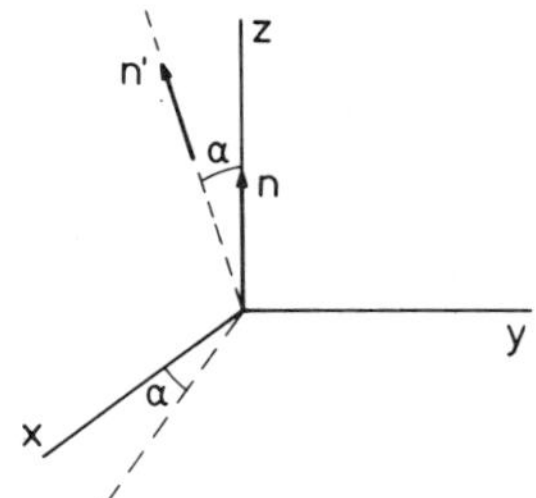

FIGURE 6.11   Definition of n and n' for the calculation of the deformation energy.

The energy density is obtained by integrating $\frac{1}{2}f$ over the volume of a sphere around the origin with radius R. We now equate the interaction energy to the macroscopic distortion energy. Choosing in Eq. (6.54) the appropriate x_i and x_j, the terms corresponding to splay, twist, bend, and second-order splay-bend are readily identified and we arrive at:

$$k_{ii} = \frac{1}{2} \int \left(\frac{\partial^2 f}{\partial \alpha^2} \right)_{\alpha=0} x_i^2 \, dV, \qquad i = 1,2,3,$$

$$k_{13}^{(2)} = \frac{1}{2} \int \left(\frac{\partial f}{\partial \alpha} \right)_{\alpha=0} x_1 x_3 \, dV. \tag{6.55}$$

In order to calculate the elastic constants from the above expressions, one has to assume a specific form for the interaction f. Nehring and Saupe (1972) have used the molecular-statistical theory of Maier and Saupe (see Chapter 1) for this purpose. In this theory, dispersion forces are considered in a dipole-dipole approximation. If we denote the induced dipoles of two molecules 1 and 2 at a mutual distance r by m_1 and m_2, their interaction energy is well known to be

$$\left[3(m_1 \cdot r)(m_2 \cdot r)/r^5 - (m_1 \cdot m_2)/r^3 \right]^2. \tag{6.56}$$

The preferred orientation of molecule 1 is taken to be given by n, that of molecule 2 by n'. In a mean field approximation, taking the surroundings of a molecule as spherically symmetric, the appropriate form of f is then:

$$f = -(ANS/r^3)^2 \left[3(n \cdot r)(n' \cdot r)/r^2 - (n \cdot n') \right]^2. \tag{6.57}$$

A is a quantity that depends on molecular properties only, N is the number of molecules per unit volume, and S is the order parameter. Evaluating Eq. (6.55) with this expression, one finds (Nehring and Saupe, 1972):

$$k_{11}:k_{22}:k_{33}:k_{13}^{(2)} = -7:11:17:-6.$$

Some of the elastic constants are predicted to be negative. However, using Eq. (6.9), the phenomenological constants are found to be positive, as they should be:

$$K_1:K_2:K_3 = 5:11:5.$$

As is to be expected with an interaction proportional to S^2, we find $K_1 = K_3$. Furthermore, K_2 is predicted to be greater than K_1, contrary to the experimental results. This is probably attributable to the fact that the molecular shape is disregarded in this calculation. For anisotropic molecules the spatial dependence of the intermolecular potential cannot be separated from the rotational dependence. Nevertheless the ratio K_3/K_1 has been related in a semi-empirical way to the anisotropy of the surroundings of a molecule, retaining the latter approximation (Priest, 1972; Gruler, 1975).

The Maier-Saupe potential can be considered as the first term of an expansion of the intermolecular potential in even-order Legendre polynomials. Priest (1973) considered the general expansion and calculated the elastic constants retaining two terms. Then the deviations from the equality $K_1 = K_3$ can be related in a simple way to the ratio $\bar{P}_4/\bar{P}_2$, where $\bar{P}_2 \equiv S$, and $\bar{P}_4$ is the average value of the fourth Legendre polynomial:

$$\bar{P}_4 = \tfrac{1}{8}\langle 35\cos^4\theta - 30\cos^2\theta + 3\rangle. \tag{6.58}$$

Introducing $\bar{K} = \tfrac{1}{3}(K_1 + K_2 + K_3)$, the result is

$$K_1/\bar{K} = 1 + \Delta - 3\Delta'\bar{P}_4/\bar{P}_2,$$

$$K_2/\bar{K} = 1 - 2\Delta - \Delta'\bar{P}_4/\bar{P}_2, \tag{6.59}$$

$$K_3/\bar{K} = 1 + \Delta + 4\Delta'\bar{P}_4/\bar{P}_2.$$

The quantities Δ and Δ' are constants depending on molecular properties. In order to calculate them, further assumptions have to be made. Representing the molecules by spherocylinders, interacting via hard core repulsions, one finds (Priest, 1973; Straley, 1973):

$$\Delta = (2R^2 - 2)/(7R^2 + 20),$$

$$\Delta' = \tfrac{9}{16}(3R^2 - 8)/(7R^2 + 20). \tag{6.60}$$

Here $R = (L - D)/D$, where L and D are the overall length and the width of the spherocylinder. As we see, for any reasonable length to width ratio Δ and Δ' are positive. Consequently, K_2 is correctly predicted to be smaller than K_1 and K_3. For positive Δ' and $\bar{P}_4/\bar{P}_2$, Eq. (6.59) always predicts $K_3 > K_1$.

Eqs. (6.59) and (6.60) can be used to predict the variation of K_3/K_1 with length/width ratio of the spherocylinder. The result is displayed in Figure 6.12 for various values of $\bar{P}_4/\bar{P}_2$. We see that for positive values of $\bar{P}_4/\bar{P}_2$ the value of K_3/K_1 tends to increase with increasing length/width ratio, in good agreement with the experimental results for rigid molecules (Leenhouts et al. 1979; compare Table 6.2).

The above results suggest that the elastic constants of molecules with flexible groups are probably "special", in the sense that there is an extra

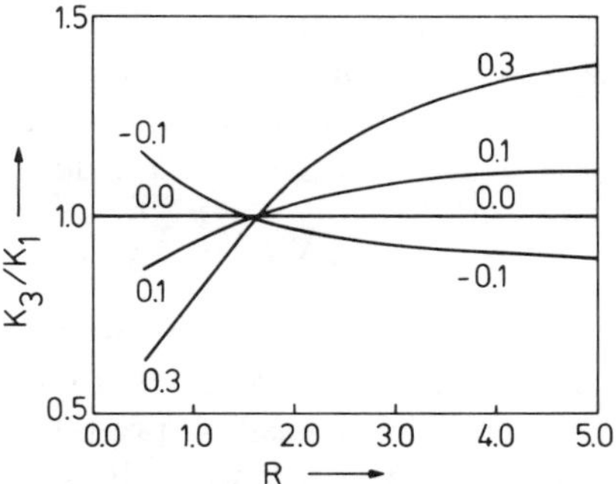

FIGURE 6.12 Variation of K_3/K_1 with the parameter R for various values of $\bar{P}_4/\bar{P}_2$ as predicted by Eqs. (6.59) and (6.60).

contribution to the elasticity from the flexible alkylene chains. A possible explanation is as follows. Compared with more rigid molecules alkyl chains may hinder a sliding of the molecules along each other. This means that neighbouring molecules will have an enhanced preference to be with their aromatic cores close to each other. Thinking of clusters of several molecules this could lead to an effective width, with L/D_{eff} possibly smaller than unity. An alternative possibility is that K_3 becomes anomalously small due to the presence of banana-shaped molecular conformations. As discussed by Gruler (1974), when banana-shaped molecules are submitted to a bend distortion, the strain can partly be relieved by changing the equilibrium distribution of molecules with one side "up" or "down". Similar to the contribution of the permanent dipole moment to the permittivity, this may lead at low frequencies to a (negative) contribution to K_3. Anyhow, evidently Eqs. (6.59) and (6.60) cannot be expected to apply to molecules with flexible groups. Hence these equations cannot be tested for MBBA, the only compound for which at present data for both $\bar{P}_4$ (Jen *et al.* 1977) and the elastic constants are available.

In conclusion we can say that our understanding of the molecular factors that influence the elastic constants of nematic liquid crystals is rather limited. Systematic experiments are still scarce, but indicate that for rigid molecules the ratio K_3/K_1 is directly related to the length/width ratio of the molecule, in agreement with theory. When flexible groups are present the situation is reversed.

We shall finish this chapter with a few remarks on the elastic properties of smectic liquid crystals. In the smectic A phase, the smectic layers can bend easily; this corresponds to a splay-type deformation of the director. Hence K_1 is not likely to be very different from its value in the nematic phase. On the other hand, twist and bend deformations are practically ruled out, as they can occur only if the layer thickness changes. The latter effect is similar to the compressibility of a liquid, and requires energies that are much higher than the energies considered here. Hence the deformation energy of the smectic A phase is in the lowest order approximation:

$$F_{\text{dist}} = \tfrac{1}{2} K_1 (\nabla \cdot \mathbf{n})^2. \tag{6.61}$$

This expression can be extended with a term corresponding to a compression of the layers (see, for example, de Gennes, 1974, Sec. 7.2).

Above a smectic-nematic phase transition pre-transitional smectic order may occur, especially when the transition is (almost) second-order. In that case, one may expect K_2 and K_3 to increase anomalously when the transition is approached from above (see, for example, Cheung *et al.* 1973; Delaye *et al.* 1973; Cladis, 1973). In Figure 6.11 these parts of the curves of elastic constants *versus* temperature have been omitted, as their inclusion obscures the "normal" temperature dependence.

CHAPTER 7

The viscosity coefficients

A nematic liquid crystal flows as easily as a conventional organic liquid consisting of similar molecules. However, when the state of alignment is considered the situation turns out to be rather complicated. In the first place, the flow depends on the angles the director makes with the flow direction and with the velocity gradient. Secondly, translational motions are coupled to inner, orientational motions of the molecules. Consequently, in most cases, the flow will disturb the alignment and cause the director to rotate. From the theoretical point of view the coupling between orientation and flow is a delicate matter. Experimentally, the direction of alignment must be controlled (for example by an external field) and/or be measured. As nematics are turbid, optical measurements of the alignment are restricted to thin samples, and conventional viscometric equipment (based on capillaries, falling spheres, rotating cylinders, *etc.*) is of little use. As a result, experiments that can be meaningfully interpreted in a quantitative way are scarce.

In this chapter our approach will be rather pragmatic. First, some basic equations derived from the hydrodynamics of isotropic liquids will be recalled (see, for example, Landau and Lifshitz, 1959). Then the extension will be made to simple shear flow of a nematic in which the orientation of the director is well controlled by an external field. Next the problem of flow alignment of the director will be discussed. Five independent coefficients with the dimension of a viscosity come into play, all of which enter into the expression for the viscous stress tensor. Experimental methods will then be considered. Finally, typical experimental results will be discussed, and we shall try to indicate how the viscosities depend on the temperature and on the order parameter. It should be emphasized that we restrict ourselves to a discussion of the viscosity coefficients. For a survey of the interesting instabilities to which the hydrodynamics of nematics may lead we refer to Dubois-Violette *et al.* (1978).

As the phenomena considered in fluid mechanics are macroscopic, the fluid is regarded as a continuous medium. Any small volume element in the fluid is always assumed to be so large that it still contains a very great number of molecules. The state of moving a fluid is described by the fluid velocity $v(r, t)$ and by any two thermodynamic quantities pertaining to the fluid, for instance the pressure $p(r, t)$ and the density $\rho(r, t)$. In the case of nematics it is necessary also to specify the director $n(r, t)$. Note that all these quantities refer to the fluid at a given point $r=(x,y,z)$ and at a given time t, *i.e.* to fixed points in space and not to fixed particles of the fluid. In the course of time, the latter move about in space.

First we shall consider the equation of continuity. The mass that flows in unit time across unit area of surface is the component of ρv normal to the surface. The total variation of ρv over the surfaces of a small volume gives the rate of decrease of the density per unit time:

$$\nabla\cdot(\rho v)=-\partial\rho/\partial t, \quad \text{or} \quad \frac{\partial(\rho v)_\alpha}{\partial x_\alpha}=-\frac{\partial\rho}{\partial t}. \tag{7.1}$$

Here we have used x_α ($\alpha=1,2,3$) for any of the variables x, y, z. We shall do so systematically in this chapter, and write sums of the type $a\cdot b=a_x b_x + a_y b_y + a_z b_z$ simply in the form $a_\alpha b_\alpha$, omitting the summation sign. A similar procedure will be used in all products involving vectors and tensors; summation over the values $1,2,3$ is always understood when a Greek suffix appears twice in any term. As an aid to the reader, we shall give at the beginning of this chapter some equations both in vector notation and for one component α. We shall restrict ourselves further to *incompressible* fluids, for which

$$\rho(r,t)=\text{constant}. \tag{7.2}$$

Consequently phenomena such as the propagation of sound waves will not be considered. With this approximation we find for Eq. (7.1)

$$\nabla\cdot v=0, \quad \text{or} \quad \frac{\partial v_\alpha}{\partial x_\alpha}=0. \tag{7.3}$$

Next we come to the equation of motion. Considering a small volume in the fluid we have

$$\rho dv/dt=f, \tag{7.4}$$

where f is the force per unit volume. The acceleration dv/dt refers to a particular piece of fluid, and is not simply equal to the acceleration $\partial v/\partial t$ at a fixed point in space. In the small time interval dt, the fluid has moved by an amount $v_x dt$ in the x direction, and by amounts $v_y dt$ and $v_z dt$ in the other directions. Hence, starting with a velocity $v(x,y,z,t)$, the velocity of the same piece of fluid at time $t+dt$ is $v(x+v_x dt, y+v_y dt, z+v_z dt, t+dt)$. From the definition of partial derivatives we have

$$v(x+v_x dt, y+v_y dt, z+v_z dt, t+dt)=v(x,y,z,t)$$

$$+(\partial v/\partial x)v_x dt+(\partial v/\partial y)v_y dt+(\partial v/\partial z)v_z dt.$$

$$\tag{7.5}$$

The acceleration is then

$$dv/dt = \partial v/\partial t + (v \cdot \nabla)v. \tag{7.6}$$

There may be various contributions to the total force f. First, the net force on the volume element due to the pressure is given by $-\nabla p$. In the presence of additional external forces that can be described by a potential ϕ (gravitation, electric or magnetic fields), there is a term $-\nabla\phi$, which we shall disregard for the moment. Finally there is the viscous term, and one finds for Eq. (7.4)

$$\rho\left[(\partial v/\partial t) + (v \cdot \nabla)v\right] = -\nabla p + f_{\text{visc}},$$

or

$$\rho\left(\frac{\partial v_\alpha}{\partial t} + v_\beta \frac{\partial v_\alpha}{\partial x_\beta}\right) = -\frac{\partial p}{\partial x_\alpha} + (f_{\text{visc}})_\alpha. \tag{7.7}$$

The force per unit area being called the stress, Eq. (7.7) can be rewritten as

$$\rho\left(\frac{\partial v_\alpha}{\partial t} + v_\beta \frac{\partial v_\alpha}{\partial x_\beta}\right) = \frac{\partial \sigma_{\beta\alpha}}{\partial x_\beta}, \tag{7.8}$$

where the elements of the stress tensor σ are defined as

$$\sigma_{\alpha\beta} = -p\delta_{\alpha\beta} + \sigma'_{\alpha\beta}. \tag{7.9}$$

Here $\delta_{\alpha\beta}$ is the Kronecker delta, and $\sigma'_{\alpha\beta}$ an element of the viscous stress tensor. For small velocity gradients, $\sigma'_{\alpha\beta}$ may be assumed to be a linear function of the spatial derivatives of the velocity, $\partial v_\alpha/\partial x_\beta$. There can be no other terms, because $\sigma'_{\alpha\beta}$ must vanish for $v = $ constant. Furthermore, if the fluid is in uniform rotation with angular velocity Ω_0, we have $v = \Omega_0 \times r$ and $\partial v_\alpha/\partial x_\beta = -\partial v_\beta/\partial x_\alpha$. Obviously, in this situation $\sigma'_{\alpha\beta}$ must also vanish. This is accomplished by taking

$$\sigma'_{\alpha\beta} = \eta\left(\frac{\partial v_\beta}{\partial x_\alpha} + \frac{\partial v_\alpha}{\partial x_\beta}\right), \tag{7.10}$$

where η is the coefficient of viscosity. The stress tensor is proportional to the symmetric part of the velocity gradient tensor. This part is often written as a tensor $\mathbf{A}$, with elements

$$A_{\alpha\beta} = A_{\beta\alpha} = \frac{1}{2}\left(\frac{\partial v_\beta}{\partial x_\alpha} + \frac{\partial v_\alpha}{\partial x_\beta}\right). \tag{7.11}$$

For incompressible fluids, the trace of the tensor $\mathbf{A}$ vanishes (see Eq. 7.3).

The antisymmetric part of the velocity-gradient tensor, which has elements

$$W_{\alpha\beta} = \frac{1}{2}\left(\frac{\partial v_\beta}{\partial x_\alpha} - \frac{\partial v_\alpha}{\partial x_\beta}\right), \tag{7.12}$$

is related to the vector ω, the vorticity of the flow, defined as

$$\omega = \tfrac{1}{2}\nabla \times v \equiv -(W_{yz}, W_{zx}, W_{xy}), \tag{7.13}$$

which describes the local angular velocity of the fluid. The decomposition of the velocity gradient tensor in a symmetric and an antisymmetric part means that, locally, the flow can be described as a superposition of an irrotational flow ($\mathbf{W}=0$) and a rotational flow ($\mathbf{A}=0$), as visualized in Figure 7.1. Insertion of Eqs. (7.9) and (7.10) in Eq. (7.8) leads to the Navier-Stokes equation for an incompressible isotropic fluid.

We shall now consider simple shear flow in a nematic liquid crystal. Let the flow be along the z axis between two parallel plates normal to the x axis. Hence the velocity gradient is along the latter axis, and we can write $v=[0,0,u(x)]$. The effective viscosity in this situation is then, according to Eq. (7.10), given by

$$\sigma_{xz} = \eta\, du/\partial x, \tag{7.14}$$

and will depend on the orientation of the director $\mathbf{n}$. The direction of $\mathbf{n}$ is specified by the angles ϕ and θ (see Figure 7.2). If we assume that the orientation of the director is fixed by external forces (for example by a

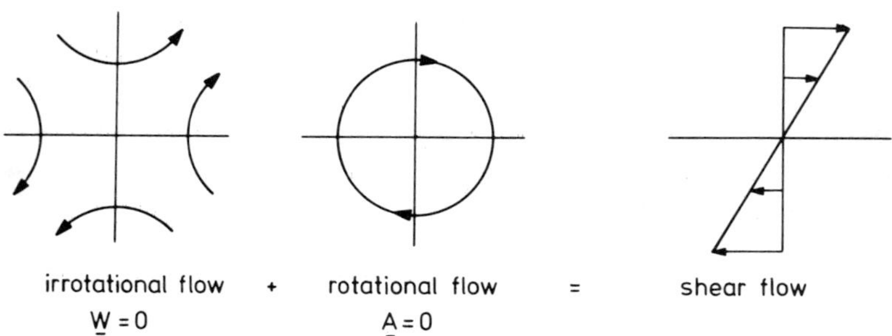

FIGURE 7.1 Irrotational and rotational flow combined to give a simple shear.

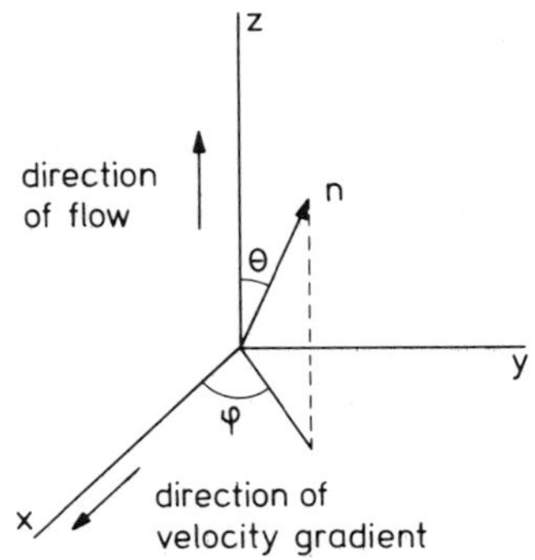

FIGURE 7.2   Definition of the orientation of the director with respect to the shear plane.

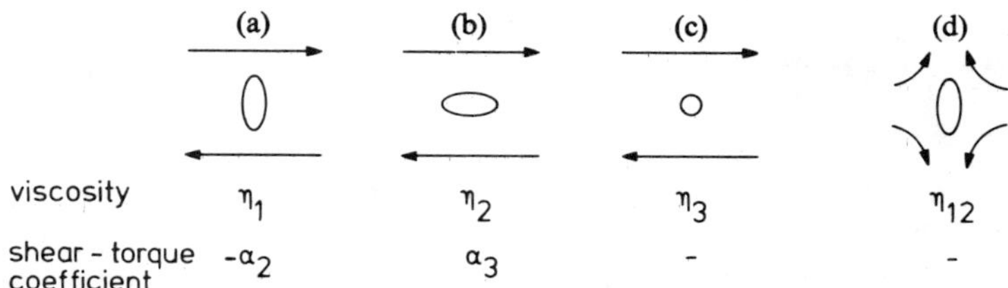

FIGURE 7.3   The viscosity coefficients of a nematic liquid crystal.

strong magnetic field), three limiting cases are immediately apparent (Figure 7.3, a, b, c):

η_1:   director parallel to velocity gradient,
$$\phi = 0°, \qquad \theta = 90°;$$
η_2:   director parallel to flow direction,
$$\phi = 0°, \qquad \theta = 0°;$$
η_3:   director normal to shear plane,
$$\phi = 90°, \qquad \theta = 90°.$$

The three coefficients η_1, η_2 and η_3 are often called Miesowicz coefficients. In the original paper by Miesowicz (1936) on PAA the definitions of η_1 and η_2 are interchanged; we have retained here the notation of Helfrich (1969, 1970).[†] In Figure 7.4 these viscosities are shown as a function of temperature for MBBA. As we see, η_3 behaves very much like a continuation of the isotropic viscosity; on the other hand η_1 and η_2 are very different.

Apart from the two possible shears depicted in Figure 7.3a and b that are antisymmetric in x and z, a stretch type of deformation is also possible

---

[†] In the literature also the notation η_a, η_b, η_c can be found, with $\eta_a = \eta_3$, $\eta_b = \eta_2$, and $\eta_c = \eta_1$.

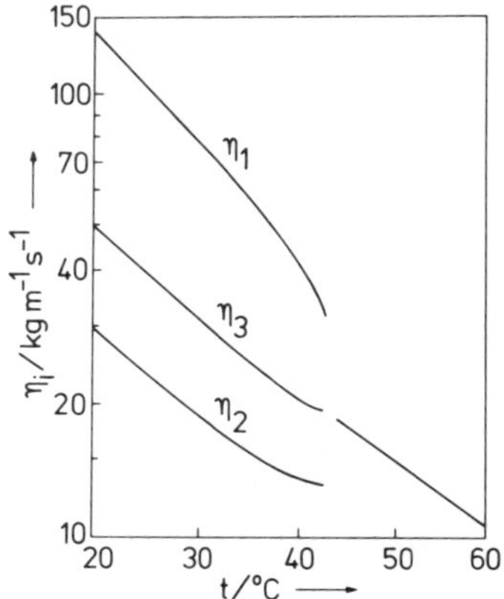

FIGURE 7.4   Miesowicz viscosities for MBBA (Gähwiller, 1973); the temperature scale is linear in $1/T$. The results for η_1 are probably too small (see p. 114).

that is symmetric in these coordinates (η_{12}, see Figure 7.3d). In a shear experiment the maximum contribution from η_{12} is found when the director is in the shear plane at an angle of $45°$ with both the flow direction and the velocity gradient. In that case

$$\eta_{45°} = \tfrac{1}{2}(\eta_1 + \eta_2) + \tfrac{1}{4}\eta_{12}. \tag{7.15}$$

For MBBA, η_{12} is much smaller than the other viscosities (Gähwiller, 1973), and the contribution from η_{12} can be disregarded. It is not clear yet whether this is true in general. The effective viscosity measured for a fixed director at arbitrary angles θ and ϕ (see Figure 7.2) is thus given by

$$\eta = \left(\eta_1 + \eta_{12}\cos^2\theta\right)\sin^2\theta\cos^2\phi + \eta_2\cos^2\theta + \eta_3\sin^2\theta\sin^2\phi. \tag{7.16}$$

So far, we have been concerned with the motion of a nematic fluid in which the orientation of the director was fixed. If we lift this restriction, an equation of motion of the director comes into play in addition. This part of the hydrodynamics illustrates the unique properties of nematic liquid crystals that have no isotropic counterpart. Experiments show that a pure rotation of the director does not necessarily involve motion of the fluid. Similar to Eq. (7.6) one has for the time derivative of the director associated with a small element of the fluid

$$d\boldsymbol{n}/dt = \partial\boldsymbol{n}/\partial t + (\boldsymbol{v}\cdot\nabla)\boldsymbol{n}. \tag{7.17}$$

With the local angular velocity of the director, given by

$$\boldsymbol{\Omega} = \boldsymbol{n}\times d\boldsymbol{n}/dt, \tag{7.18}$$

one thus obtains the equation of motion

$$I d\Omega/dt = \Gamma_F + \Gamma_{\text{visc}}.$$ (7.19)

The left-hand side represents the inertial term; I is the moment of inertia per unit volume. The first term on the right-hand side is the torque per unit volume on the director due to the elastic (and possibly magnetic and electric) forces,

$$\Gamma_F = n \times h,$$ (7.20)

where h is the molecular field (see Eq. 6.17). The second term is a torque on the director due to frictional forces. As before, we assume that this term is a linear function of the velocity gradients (see Eq. 7.10) and, of course, of the motion of the director relative to its surroundings. This latter effect is not simply given by dn/dt. If the fluid rotates as a whole with angular velocity ω, the viscous torque will depend on

$$N = (\Omega - \omega) \times n = dn/dt - \omega \times n.$$ (7.21a)

Using Eq. (7.13), this can also be written as

$$N_\alpha = \frac{dn_\alpha}{dt} - W_{\alpha\beta} n_\beta.$$ (7.21b)

Two linearly independent terms can be formed that are linear in the derivatives and have the correct physical meaning of an axial vector normal to n (Saupe, 1973):

$$n \times N, \qquad n \times \mathbf{A} \cdot n.$$

The viscous torque is obtained as the sum of these terms, each multiplied by an appropriate coefficient (shear torque coefficient)

$$\Gamma_{\text{visc}} = -\gamma_1 n \times N - \gamma_2 n \times \mathbf{A} \cdot n.$$ (7.22)

At low frequencies, the inertial term in Eq. (7.19) is much smaller than the elastic and the viscous torque. Then Eq. (7.19) is referred to as the balance of torques, which reads, after inserting the results from Eqs. (7.20) and (7.22):

$$n \times h = \gamma_1 n \times N + \gamma_2 n \times \mathbf{A} \cdot n.$$ (7.23)

Now let us go back to the simple shear experiment of Figure 7.2, with

$$v = [0, 0, u(x)],$$

$$n = (\sin\theta \cos\phi, \sin\theta \sin\phi, \cos\theta).$$

In that case we find

$$A_{xz} = W_{xz} = -W_{zx} = \tfrac{1}{2}\,du/dx,$$

$$N_z = \omega_y n_x = -A_{xz} n_x,$$

$$N_x = -\omega_y n_z = A_{xz} n_z.$$

If $\mathbf{n}$ is perpendicular to the shear plane ($\theta = \phi = 90°$), there is no torque, and $\Gamma_{\mathrm{visc}} = 0$. If we consider the situation in the shear plane ($\phi = 0°$), where $\mathbf{n} = (\sin\theta, 0, \cos\theta)$, one finds a torque along the y axis given by:

$$\Gamma_{\mathrm{visc}} = -\gamma_1(n_z N_x - n_x N_z) - \gamma_2(n_z n_\mu A_{\mu x} - n_x n_\mu A_{\mu z})$$

$$= -\tfrac{1}{2}(du/dx)\left[\gamma_1 + \gamma_2(\cos^2\theta - \sin^2\theta)\right]. \tag{7.24}$$

The shear torque vanishes and a stable orientation exists for $\mathbf{n}$ in the shear plane at an angle θ_0 given by

$$\cos 2\theta_0 = -\gamma_1/\gamma_2. \tag{7.25}$$

The angle θ_0 is called the flow alignment angle. Apart from a small boundary layer, the elastic torque can be disregarded, and the result is the director pattern of Figure 7.5. The flow alignment angle is independent of the shear rate, and exists only if $|\gamma_1/\gamma_2| < 1$.

The hydrodynamic properties of a nematic liquid crystal are often described more formally. Similar to Eq. (7.9) a stress tensor, taking account of the orientation of the director, can be written as

$$\sigma_{\alpha\beta} = -p\delta_{\alpha\beta} - \frac{\partial F}{\partial g_{\alpha\gamma}} g_{\beta\gamma} + \sigma'_{\alpha\beta}, \tag{7.26}$$

where $g_{\alpha\beta} = \partial n_\beta/\partial x_\alpha$. Compared with the isotropic case, there is an additional term, the stress connected with the distortion free energy density F. For small deformations, this term plays no role, and usually it can be disregarded. The important differences come in with the viscous stress tensor $\sigma'_{\alpha\beta}$. One expects its components to depend linearly on both $\mathbf{A}$ (see Eq. 7.11) and $\mathbf{N}$ (see Eq. 7.21). Furthermore, as we expect $\sigma'(-\mathbf{n}) = \sigma'(\mathbf{n})$,

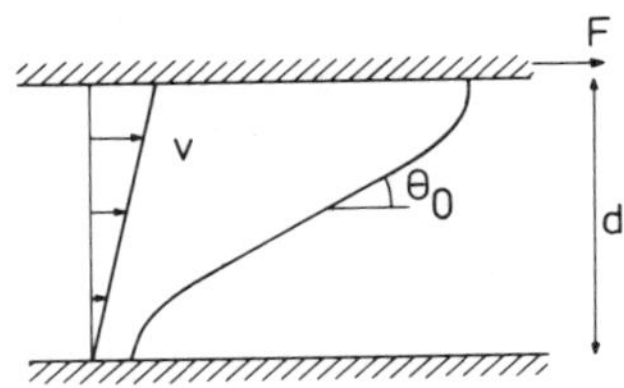

FIGURE 7.5   Flow alignment in a simple shear experiment of a nematic fluid.

the stress tensor must be an even function of n. The most general form for an incompressible nematic, satisfying these requirements is:

$$\sigma'_{\alpha\beta} = \alpha_4 A_{\alpha\beta} + \alpha_1 n_\alpha n_\beta n_\mu n_\rho A_{\mu\rho}$$

$$+ \alpha_2 n_\alpha N_\beta + \alpha_3 n_\beta N_\alpha + \alpha_5 n_\alpha n_\mu A_{\mu\beta} + \alpha_6 n_\beta n_\mu A_{\mu\alpha}. \tag{7.27}$$

Note that the tensor is not symmetric. There are six coefficients with the dimension of a viscosity (Leslie coefficients). Parodi (1970) has shown that from Onsager's reciprocal relations, which reflect the time reversal invariance of the microscopic motions, one can derive

$$\alpha_6 = \alpha_2 + \alpha_3 + \alpha_5. \tag{7.28}$$

Hence only five coefficients are independent.

The formal theory connected with Eqs. (7.27) and (7.28) is derived from considerations of the dissipation, or as it is called, the entropy source, due to all friction processes in the fluid. For a discussion we refer to de Gennes' book (1974) or the review paper by Leslie (1979), where further references to alternative formulations can also be found. For the present purpose it suffices to note that the fact that the entropy source must be positive imposes some restrictions on the viscosity coefficients. Considering Eq. (7.27) in more detail we see that in the isotropic phase only α_4 survives; comparison with Eqs. (7.10) and (7.11) shows that

$$\eta_{is} = \tfrac{1}{2}\alpha_4. \tag{7.29}$$

Now let us return to the shear experiment of Figure 7.2. In that case one easily finds for the stress tensor

$$\sigma_{xz} = \tfrac{1}{2}(du/dx)\big[(2\alpha_1 \cos^2\theta - \alpha_2 + \alpha_5)\sin^2\theta \cos^2\phi + (\alpha_3 + \alpha_6)\cos^2\theta + \alpha_4\big].$$

$$\tag{7.30}$$

This expression is equivalent to the equation for the effective viscosity, Eq. (7.16), with

$$\eta_1 = \tfrac{1}{2}(-\alpha_2 + \alpha_4 + \alpha_5),$$

$$\eta_2 = \tfrac{1}{2}(\alpha_3 + \alpha_4 + \alpha_6),$$

$$\eta_3 = \tfrac{1}{2}\alpha_4. \tag{7.31}$$

$$\eta_{12} = \alpha_1.$$

The fact that η_1, η_2 and η_3 are necessarily positive, imposes some restrictions on the Leslie coefficients. With these relations the Onsager-Parodi relation can be written as

$$\eta_2 - \eta_1 = \alpha_3 + \alpha_2. \tag{7.32}$$

The shear torque acting on the director can be expressed in terms of the antisymmetric part of $\sigma'_{\alpha\beta}$:

$$(\Gamma_{\text{visc}})_y = \sigma'_{zx} - \sigma'_{xz}, \quad etc. \tag{7.33}$$

For the shear geometry of Figure 7.2 with $\phi = 0$, this leads to

$$\Gamma_{\text{visc}} = (\alpha_5 - \alpha_6)(n_z n_\mu A_{\mu x} - n_x n_\mu A_{\mu z}) + (\alpha_2 - \alpha_3)(n_z N_x - n_x N_z)$$

$$= (du/dx)(\alpha_2 \sin^2 \theta - \alpha_3 \cos^2 \theta). \tag{7.34}$$

We see that $-\alpha_2$ and α_3 are the coefficients relating the torque to the shear rate when the director is parallel to the velocity gradient (configuration 1), or to the flow direction (configuration 2), respectively (see Figure 7.3)[†]. This gives a simple interpretation to the Onsager-Parodi relation when written in the form of Eq. (7.32). It says that the shear stress in configuration 2 produced by the shear in configuration 1 equals the shear stress in configuration 1 produced by shear flow of the same strength in configuration 2 (Helfrich, 1970). Comparison of Eq. (7.34) with Eq. (7.24) reveals that

$$\gamma_1 = \alpha_3 - \alpha_2,$$
$$\gamma_2 = \alpha_6 - \alpha_5. \tag{7.35}$$

Using Eq. (7.28) an alternative expression for the flow alignment angle can now be given (compare Eq. 7.25):

$$\tan^2 \theta_0 = \alpha_3 / \alpha_2 = (\gamma_2 + \gamma_1)/(\gamma_2 - \gamma_1). \tag{7.36}$$

In nematics consisting of rod-like molecules, for which $\eta_1 > \eta_2$, the condition of increasing entropy leads to $\alpha_2 < 0$. Hence flow alignment occurs only if $\alpha_3 < 0$. Quantitative results of measurements of θ_0 are scarce but some examples are given in Figure 7.6. More qualitative results indicate that, at least far from T_{NI}, θ_0 is often rather small ($|\alpha_3|$ is very small). Hence flow experiments in which the orientation of the director is not controlled will often approximately give η_2 (**n** parallel to the flow direction). It turns out that in some cases, however, $\alpha_3 > 0$, to which situation we shall return later.

Various methods can be used to measure the nematic viscosities. We shall now discuss some of the most important ones in more detail.

[†] In Helfrich's notation $-\alpha_2 = \kappa_1$ and $\alpha_3 = \kappa_2$.

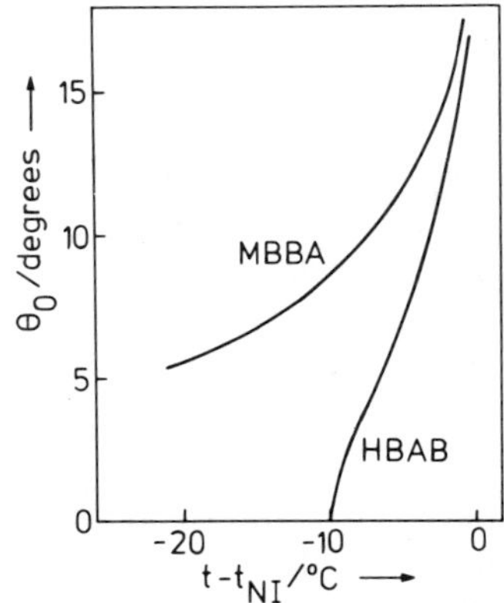

FIGURE 7.6   Flow alignment angle for some nematics (Gähwiller, 1972).

(a)   Shear flow in a magnetic field

Measurements of shear flow using a strong magnetic field to control the orientation of the director are already quite old (Miesowicz, 1936; Tsvetkov and Michailov, 1938) and have been taken up again more recently (Gähwiller, 1973). Of course, the magnetic torque has to be much larger than the torque exerted by the shear flow. For small shear rates, this situation can easily be reached. Choosing the appropriate geometries one can thus determine η_1, η_2, η_3 and η_{12}. Hence, the additional measurement of θ_0 is in principle sufficient to determine a complete set of viscosity coefficients. Two methods have been used for the measurement of θ_0.

i) If a rectangular capillary is made using two glass plates with conductive electrodes, the capacitance can be measured with a strong magnetic field parallel and perpendicular to the flow and without a magnetic field. Provided the flow is uniform, and the boundary layers can be disregarded, these results suffice to calculate θ_0 (Marinin and Tsvetkov, 1939; Meiboom and Hewitt, 1973). The drawback of the method is that it is usually difficult to monitor the flow optically at the same time, which is necessary to check the alignment.

ii) The change in birefringence can be measured as a function of the flow (Gähwiller, 1973). Let us look again at the situation of Figure 7.2, with a rectangular capillary formed using two glass plates at $x = \pm \frac{1}{2}d$. When the nematic is at rest, with the director parallel to the z axis (either through appropriate boundary conditions or as a result of a strong magnetic field parallel to the z axis), the optical path difference between the ordinary and extraordinary component of polarized light shining perpendicular to the glass plates is $(n_e - n_o)d$. With shear flow this path difference becomes

$$\int_{-d/2}^{d/2} \left[n(\theta) - n_o \right] dx,$$

where $n(\theta)$ is given by Eq. (6.27). Once the flow alignment is established, θ varies from zero to $+\theta_0$ (for $x<0$) and $-\theta_0$ (for $x>0$), respectively. The change in optical path difference for small values of θ_0 is then:

$$-\tfrac{1}{2} n_e d \left(n_e^2 / n_o^2 - 1 \right) \tan^2 \theta_0. \tag{7.37}$$

This difference is independent of the shear rate, provided that the thickness of the transition layers at the glass plates and at $x=0$ (where $\theta(x)$ varies from $+\theta_0$ to $-\theta_0$) is negligible compared with the plate separation. The optical path difference can be measured accurately by recording the oscillations of the transmitted light on application of the shear. In this way, $\tan^2 \theta_0$ and thus α_3/α_2 are determined. By application of a magnetic field, the change in birefringence can be measured as a function of the field. In this way $\alpha_3/\Delta\chi$ and $\alpha_2/\Delta\chi$ can be measured separately (Gähwiller, 1973).

(b)   Rotating field method

We consider the situation of Figure 7.7, where a uniform magnetic field of induction $\boldsymbol{B}$ is applied to a cylindrical nematic sample. To obtain a situation of minimum energy, the director will like to be parallel to the field. If the field is slowly rotated the director will follow the field, but with a certain phase lag. This phase lag is such that the frictional torque and the magnetic torque balance, and this allows a measurement of γ_1 to be made.

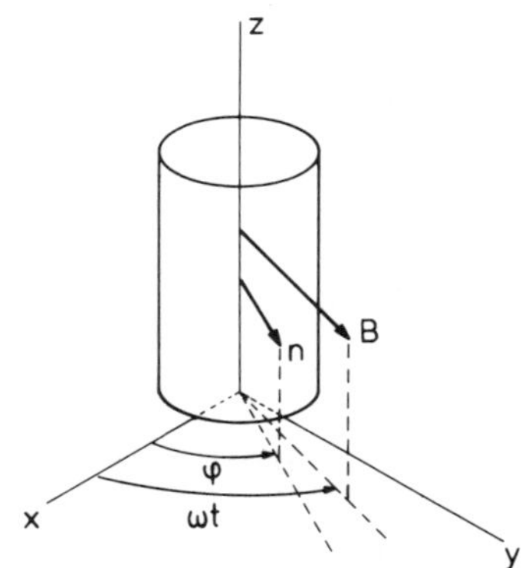

FIGURE 7.7   Nematic sample in a rotating magnetic field.

Referring to Figure 7.7, we have

$$\mathbf{B} = (B_0 \cos \omega t, B_0 \sin \omega t, 0),$$

$$\mathbf{n} = (\cos \phi, \sin \phi, 0),$$

leading to $\mathbf{N} = d\mathbf{n}/dt = (-\sin \phi, \cos \phi, 0)d\phi/dt$. From Eq. (7.22), we obtain a shear torque in the z direction

$$\Gamma_{\text{visc}} = -\gamma_1 d\phi/dt \tag{7.38}$$

Disregarding elastic effects, this torque, according to Eq. (7.23), should be balanced by the magnetic torque, leading to

$$\gamma_1 \frac{d\phi}{dt} = \tfrac{1}{2}\mu_0^{-1}\Delta\chi B_0^2 \sin 2(\omega t - \phi). \tag{7.39}$$

The solutions of this differential equation depend on the value of

$$\omega_0 = \left(\mu_0^{-1}\Delta\chi/2\gamma_1\right)B_0^2. \tag{7.40}$$

For $\omega < \omega_0$, a steady state is reached for which there is a constant phase lag between director and field given by

$$\tan(\omega t - \phi) = \omega_0/\omega - \left(\omega_0^2/\omega^2 - 1\right)^{\frac{1}{2}}. \tag{7.41}$$

For solutions in the case $\omega > \omega_0$, we refer to Gasparoux and Prost (1971). If we discount any contribution from the bottom of the cell, and assume that the torque exerted on the nematic is fully transmitted to the sample container, the torque on the container for $\omega < \omega_0$ is

$$\Gamma_{\text{cont}} = V\gamma_1\omega, \tag{7.42}$$

where V is the volume of the sample. Γ_{cont} can be measured by suspending the sample on a quartz fibre in a magnetic field and measuring the torsion on application of the field. The slope of the straight line Γ_{cont}/V *versus* ω gives directly γ_1.

As discussed by de Gennes (1974), there are some fundamental problems in this experiment connected with the anchoring of the molecules at the walls. Because of this anchoring, a rotation of $\mathbf{B}$ does not give a real steady-state region, but a twist between both surfaces that would increase linearly with time. This twist must relax *via* disclination loops. In bulk samples at low frequencies, the fraction of the sample in which the orientation is perturbed by the disclination lines is probably small, which explains the success of the method.

(c) Dynamics of the Frederiks transition

Information about the viscous behaviour of a nematic can be obtained by studying the response to a sudden change of an applied field. We shall restrict our attention to the situation of a magnetic field which is stronger than the threshold field of a Frederiks transition, and which is suddenly removed. First we consider the second case of Figure 6.2, involving a pure twist. In this situation there is no hydrodynamic flow—the molecules rotate without any translational motion—and this simplifies the analysis considerably. If the applied field B is not much stronger than the threshold field B_c, the twist angle $\phi_0(z)$ is small. With the substrates at $z = \pm \frac{1}{2}d$, where necessarily $\phi = 0$, we have to a good approximation

$$\phi_0(z) = \phi_m \cos(\pi z / d), \tag{7.43}$$

where ϕ_m is the maximum distortion at $z = 0$. If the field is removed, the viscous torque is equal to the elastic torque (see Eq. 7.23), which gives the dynamic equation for $\phi(z, t)$:

$$K_2 \partial^2 \phi / \partial z^2 = \gamma_1 \partial \phi / \phi t, \tag{7.44}$$

with the initial condition that $\phi = \phi_0(z)$ at $t = 0$. The solution is

$$\phi(z, t) = \phi_0(z) \exp(-t / \tau), \tag{7.45}$$

with a relaxation time

$$\tau = \gamma_1 d^2 / (\pi^2 K_2) = \mu_0 \gamma_1 / (\Delta \chi B_c^2). \tag{7.46}$$

Plotting τ versus B_c^{-2} for various thicknesses results in a straight line whose slope is given by $\gamma_1 / \Delta \chi$. Note that the simple treatment given here cannot be expected to apply to the region $B \gg B_c$, because of the assumption involved in Eq. (7.43).

The situation for the splay geometry is more complicated. In this case, rotation of the director causes hydrodynamic flow: the gradient of the angular velocity of n produces a backflow motion of the fluid, giving a frictional torque. From the balance of torques and the equation of motion a solution for τ can be obtained that depends in a rather complicated way on B / B_c. The result can be written in the form of Eq. (7.46), however, with γ_1 replaced by an effective viscosity γ_1^* (Pieranski *et al.* 1973). In practice γ_1^* may be some 10% smaller than γ_1. There is some special interest in the case $B / B_c \to \infty$. Then γ_1^* reduces to

$$\eta_{\mathrm{splay}} = \gamma_1 - \alpha_3^2 / \eta_2. \tag{7.47}$$

In the bend geometry a similar relation applies, with as limiting case

$$\eta_{\text{bend}} = \gamma_1 - \alpha_2^2/\eta_1. \tag{7.48}$$

As $|\alpha_3| \ll |\alpha_2|$, the corrections to γ_1 are usually small for the splay geometry, but they can be quite important if we study the deformation of a homeotropic sample.

(d) Light scattering

As already discussed in the previous chapter, in a nematic liquid crystal the long wavelength fluctuations of the optical axis give rise to a large scattering of light. These fluctuations $\delta n(r,t) = n(r,t) - n_0$ are excited at certain volume elements and subsequently relax towards zero. This dynamic process may be described by the hydrodynamic equations. Experimentally, these time-dependent fluctuations can be observed, because they lead to a frequency modulation of the scattered light that is measurable with present optical techniques (Orsay liquid crystal group, 1969; Van Eck and Westera, 1977). We shall not discuss these effects at length, but simply summarize the type of information that can be obtained.

The scattered amplitude, for a given scattering wave vector q, depends on two independent Fourier components $n_1(q)$ and $n_2(q)$ of the fluctuation δn (see Figure 6.5). We recall that n_1 describes in general a mixed deformation involving splay and bend, while n_2 involves twist and bend. Each of them may be analysed separately by a suitable choice of polarizations. For each mode there is a restoring force given by

$$K_\alpha(q) = K_\alpha q_\perp^2 + K_3 q_\parallel^2, \qquad \alpha = 1, 2. \tag{7.49}$$

Furthermore, it turns out that the experimental observations of the line width as a function of the wave vector can be described in a model involving a purely visco-elastic type of relaxation:

$$\frac{\partial}{\partial t} n_\alpha(q) = -\tau_\alpha^{-1} n_\alpha(q), \qquad \alpha = 1, 2. \tag{7.50}$$

In terms of the power spectrum, this gives the experimentally observed, single Lorentzian line of a width that is the inverse of the time τ. The time τ is given by

$$\tau_\alpha^{-1}(q) = \frac{K_\alpha(q)}{\eta_\alpha^{\text{eff}}(q)}, \qquad \alpha = 1, 2, \tag{7.51}$$

where an effective viscosity has been introduced:

$$\eta_1^{\text{eff}}(\boldsymbol{q}) = \gamma_1 - \frac{\left(q_\perp^2 \alpha_3 - q_\parallel^2 \alpha_2\right)^2}{q_\perp^4 \eta_2 + q_\perp^2 q_\parallel^2 (\alpha_1 + \alpha_3 + \alpha_4 + \alpha_5) + q_\parallel^2 \eta_1},$$

$$\eta_2^{\text{eff}}(\boldsymbol{q}) = \gamma_1 - \frac{\alpha_2^2 q_\parallel^2}{q_\perp^2 \eta_3 + q_\parallel^2 \eta_1}. \tag{7.52}$$

Two situations are of special interest. If $\boldsymbol{q}$ is along the director, both modes reduce to a pure bend, and $\eta_1^{\text{eff}} = \eta_2^{\text{eff}} = \eta_{\text{bend}}$ (Eq. 7.48). If $\boldsymbol{q}$ is normal to the director, mode (1) becomes a pure splay ($\eta_1^{\text{eff}} = \eta_{\text{splay}}$, Eq. 7.47), and mode (2) a pure twist ($\eta_2^{\text{eff}} = \gamma_1$). With this method, a ratio between an elastic constant and a viscosity is obtained. One quantity must be obtained by a different method in order to obtain a value for the other.

(e) Ultrasonic shear waves

The action of ultrasound may be used to investigate the viscous behaviour of a nematic liquid crystal. The classical experiment is due to Martinoty and Candau (1971), who measured the reflection coefficient of an ultrasonic shear wave at a solid-nematic interface. The experimental set-up is shown in Figure 7.8. The reflection coefficient is given by

$$r = (Z_s - Z_n)/(Z_s + Z_n), \tag{7.53}$$

where Z_s and Z_n are the complex impedances of the solid and the nematic, respectively. Z_s being known, the measurement determines the mechanical impedance $Z_n = R_n + iX_n$. With this impedance is associated a dynamic viscosity through the relation

$$\tilde{\eta} = 2R_n X_n/\rho\omega = 2R_n^2/\rho\omega, \tag{7.54}$$

where ω is the frequency, ρ is the density, and where the experimental result $R_n = X_n$ has been used. The actual expression for $\tilde{\eta}$ depends on the orientation of the director at the interface.

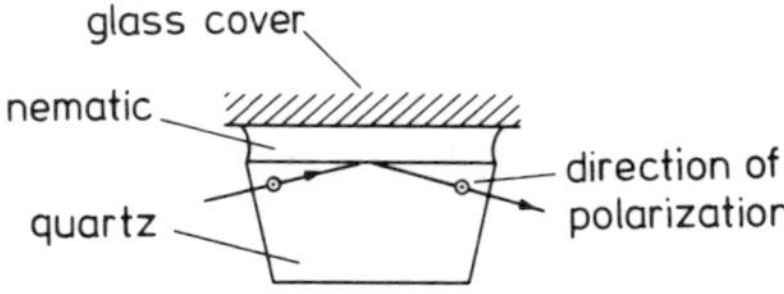

FIGURE 7.8 Experimental set-up for the reflection of ultrasonic shear waves at a solid-nematic interface.

We shall not give the calculation of $\tilde{\eta}$ in detail for the various situations, but simply state the results. Let us take a coordinate system compatible with Figure 7.2, with the propagation direction along the x axis and the displacement (flow) along the z axis. The difference compared with the Miesowicz experiment is that because of the high frequencies involved (say 50 MHz), the penetration thickness is only a few microns. Hence there is a strong velocity gradient in the propagation direction (x direction). The orientation of the director at the interface is controlled by the boundary conditions. However, in the absence of an orienting magnetic field, the ultrasonic shear will in general impose a non-negligible tilt. For that reason $\tilde{\eta}_1$ and $\tilde{\eta}_2$ will differ from the Miesowicz viscosities η_1 and η_2. If the director is normal to the shear plane, of course, no tilt is induced. The results are:

$\boldsymbol{n}$ parallel to velocity gradient: $\tilde{\eta}_1 = \eta_1 + \frac{1}{2}\alpha_2(1 - \gamma_2/\gamma_1)$;

$\boldsymbol{n}$ parallel to flow direction: $\quad \tilde{\eta}_2 = \eta_2 - \frac{1}{2}\alpha_3(1 + \gamma_2/\gamma_1)$;

$\boldsymbol{n}$ normal to shear plane $\quad\quad \tilde{\eta}_3 = \eta_3$.

With Eqs. (7.31) and (7.35), it is easily shown that $\tilde{\eta}_1 = \tilde{\eta}_2$, which result is independent of the validity of the Onsager-Parodi relation, Eq. (7.28).

Ultrasonic experiments have been extended to include the attenuation as a function of the angle between the director and the wave vector (Bacri, 1974). The shear impedance of some smectic liquid crystals has also been studied (Kiry and Martinoty, 1977).

Now we come to a discussion of some typical values for the viscosity coefficients. In that context it is disappointing that a complete set of coefficients can be given for only very few compounds. In Table 7.1 we show such a set for PAA at 122°C. To do so we have to go back to the old

TABLE 7.1.
Viscosities of PAA at 122°C (10^{-3} kgm^{-1}s^{-1})

Experimental	Calculated	
$\eta_1 = 9.2 \pm 0.4$[a]	$\alpha_1 = 4 \pm 4$	$\eta_2 = 2.1 \pm 0.1$
$\eta_{\theta_0} = 2.43 \pm 0.05$[a]	$\alpha_2 = -6.9 \pm 0.2$	$\gamma_2 = -7.1 \pm 0.3$
$\eta_3 = 3.4 \pm 0.4$[a]	$\alpha_3 = -0.2 \pm 0.1$	
$\theta_0 = 9° \pm 1°$[b]	$\alpha_4 = 6.8 \pm 0.8$	
$\gamma_1 = 6.8 \pm 0.2$[c]	$\alpha_5 = 5 \pm 1$	
	$\alpha_6 = -2 \pm 1$	

[a]Miesowicz (1936); [b]Marinin and Tsvetkov (1939); [c]Gasparoux and Prost (1971)

shear flow measurements of Miesowicz (1936). As noted by Tseng *et al.* (1972) in a critical discussion of these data, Miesowicz was not able to apply a magnetic field parallel to the flow. His value of η_2 is given assuming $\theta_0 = 0°$. Tsvetkov and Michailov (1938) applied a magnetic field in the direction of the flow and found a decrease in the viscosity of 16%, corresponding to $\theta_0 \approx 9°$. This value for θ_0 was confirmed by capacitance measurements with and without a field (Marinin and Tsvetkov, 1939). Together with a recent determination of γ_1, and taking the Onsager-Parodi relation into account, this suffices to calculate α_1 to α_6. As we see from Table 7.1, the accuracy with which α_1 is obtained is rather poor. It has been argued that $\alpha_1 \approx 0$, and this is quite possible within the experimental accuracy. The results in Table 7.1 agree reasonably well with data from light scattering experiments (Orsay liquid crystal group, 1971). The latter results are probably somewhat less accurate.

For MBBA, a complete set of viscosity coefficients has been given by Gähwiller (1973). From the data in Figure 7.4, the Onsager-Parodi relation, and the measurement of θ_0, one can calculate α_2 and α_3 (see Figure 7.9). A measurement of $\eta_{45°}$, see Eq. (7.15), gives the value of η_{12}, which completes the results. As a check on these data, de Jeu (1978b) collected results for γ_1 obtained from various sources using different methods (see Figure 7.10). These latter results indicate (at 25°C) a value of $\gamma_1 = 95 \times 10^{-3}$ kgm^{-1}s^{-1}, rather larger than Gähwiller's result. Summerford *et al.* (1975) have argued that Gähwiller's result for η_1 is too small, because in this particular configuration (field parallel to velocity gradient) the magnetic field was not strong enough to prevent flow alignment at the shear rates used. This argument does not hold for η_2 and η_3. Indeed Gähwiller's results for η_2 and η_3 of MBBA at room temperature, are in good agreement with the values obtained from the reflection of ultrasound. In Table 7.2 we summarize the results for the viscosities of MBBA, not including the experimental value

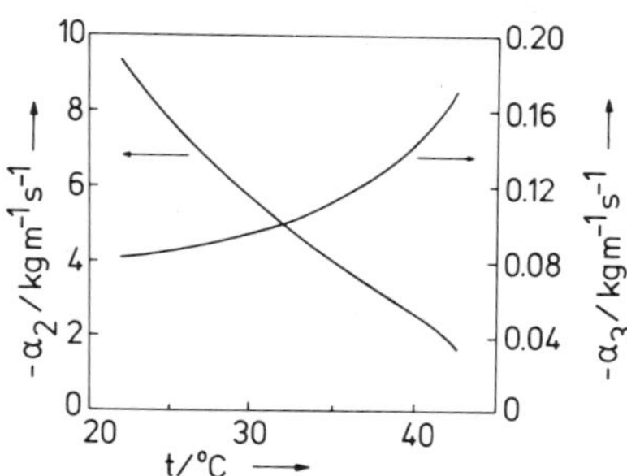

FIGURE 7.9   Shear torque coefficients for MBBA (Gähwiller, 1973).

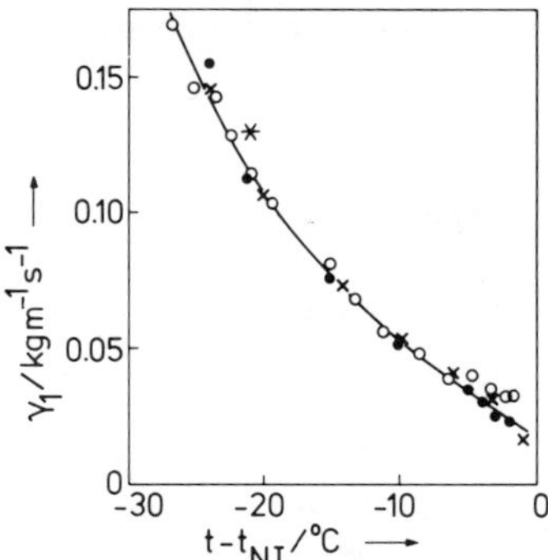

FIGURE 7.10   Rotational viscosity γ_1 for MBBA from various methods:
●✕: rotating magnetic field (Gasparoux and Prost, 1971; Heppke and Schneider, 1972),
*: motion of walls (Léger, 1972), ○: light scattering (Shaya and Yu, 1975; calculated as indicated by de Jeu, 1978b).

TABLE 7.2.
Viscosities of MBBA at 25°C ($t_{NI} - t = 18°C$)
in $10^{-3}\,\mathrm{kgm^{-1}s^{-1}}$

Experimental	Calculated	
$\eta_2 = 23.8 \pm 0.3^a$	$\alpha_1 = 6.5 \pm 4$	$\eta_1 = 121 \pm 4$
$\eta_3 = 41.6 \pm 0.7^a$	$\alpha_2 = -96 \pm 3$	$\gamma_2 = -97 \pm 3$
$\eta_{12} = 6.5 \pm 4^a$	$\alpha_3 = -1.1 \pm 0.2$	
$\theta_0 = 6° \pm 1^{oa}$	$\alpha_4 = 83.2 \pm 1.5$	
$\gamma_1 = 95 \pm 3^b$	$\alpha_5 = 63 \pm 12$	
	$\alpha_6 = -34 \pm 2$	

^{a}Gähwiller (1973); ^{b}from Figure 7.10

for η_1, which quantity has been calculated with the Onsager-Parodi relation, Eq. (7.32). Now a check is possible with results from light scattering experiments. From Table 7.2 we calculate

$$\eta_{\mathrm{bend}} = \gamma_1 - \alpha_2^2/\eta_1 = 19 \times 10^{-3}\,\mathrm{kgm^{-1}s^{-1}}.$$

Experimental results for this quantity range from 19×10^{-3} to 21×10^{-3} $\mathrm{kgm^{-1}s^{-1}}$ (Haller, 1972; Martinand and Durand, 1972). Although these latter results are not very accurate, because they depend on the value chosen for K_3, we see that the agreement is excellent. It should be said, however, that the value of η_{bend} is not very sensitive to variations in the value of γ_1.

Only for a few other compounds are extensive viscosity data available. Among them are:

$$C_6H_{13}O-\bigcirc-CH=N-\bigcirc-CN \qquad \text{and} \qquad NC-\bigcirc-CH=N-\bigcirc-OC_8H_{17}$$

HBAB   CBOOA

The viscosities of HBAB have been reported by Gähwiller (1973), and those of CBOOA by Kim *et al.* (1976). Both compounds show the peculiarity that, below a certain temperature α_3 changes sign and becomes positive. This means that there is no longer any real solution to Eq. (7.36), and that no flow alignment can occur (see Figure 7.6 for HBAB below 92°C). The original observation of this effect with HBAB has been contested by Meiboom and Hewitt (1973), but later has been shown to be correct (Pieranski and Guyon, 1974; White *et al.* 1977). For $\alpha_3 > 0$, the director tumbles in a way that depends on the pre-tumbling velocity field in the sample (Cladis and Torza, 1975).

Helfrich (1969) has given a molecular theory of flow alignment in nematic liquid crystals. It is based on the rod-like shape of the molecules, which are considered as ellipsoids of revolution. As is readily seen, no torque can be exerted on a sphere; since the forces acting on any surface element are normal, they always point to the centre and going through the centre of gravity, they do not provide a lever arm. On the other hand, a torque can arise if the body is an ellipsoid, because the forces due to pressure do not in general point towards the centre of gravity. Thus it is inherent in the model that there can be no shear torque if the molecules are perpendicular to the shear plane, as the cross-sections in that plane are circular. Two severe simplifications are made to calculate the torques in the other situations: (a) the ellipsoids are assumed to be equally and rigidly oriented (corresponding to $S = 1$); (b) the nematic liquid is treated as a gas, so that the hard ellipsoids interact only by elastic collisions. In a real nematic liquid crystal, each molecule constantly feels the attractive potential of its neighbours. In violation of assumption (b), the resulting short-range order may involve preferential points of contact and altered collision rates. Both effects, while changing the calculation of the shear torque coefficients, should not influence their ratio. Therefore the most important result of the model is the prediction

$$\alpha_3/\alpha_2 = (b/a)^2, \tag{7.55}$$

where a and b are the long and the short axes of the ellipsoid, respectively. Indeed Eq. (7.55) is retained in a second model calculation (Helfrich, 1970) in terms of the theory of dense fluids, in which the still equally and rigidly oriented molecules interact through a smooth ellipsoidal two-body potential. For molecules like PAA and MBBA the ratio a/b is of the order of 5, giving $\alpha_3/\alpha_2 \approx 0.04$. Comparison with Tables 7.1 and 7.2 reveals that this is in excellent agreement with experiment.

The difference in response to shear flow of HBAB ($\alpha_3 > 0$) and MBBA ($\alpha_3 < 0$) is particularly striking if we consider that the two molecules are similar both in size and in chemical structure. It has been suggested that the dipolar interaction associated with the cyano groups in HBAB is responsible for the difference (Gähwiller, 1972). As already noted in Chapter 5, we can conclude from dielectric measurements that when strongly polar groups are present, an appreciable number of the molecules will be associated with the polar heads antiparallel. In the light of the above model, one might consider this in a loose way as a "bumpy" contour, in contrast to the ellipsoid which can be used to represent a molecule of PAA or MBBA. Placed in a shear field, such an associated entity is likely to feel a torque of positive sign even if it is oriented parallel to the direction of flow. Of course, such an effect can be expected especially at low temperatures, where the association is stronger.

Helfrich's model may be used to establish a relationship between the shear torque coefficient and the viscosity under the same type of shear. If the short-range order were such that the ellipsoids collided only near their ends and around their centre circles (we still assume rigid and equal molecular orientation), one would have $-\alpha_2 \approx \eta_1$. With a uniform distribution of the collisions over the body surface one would expect $|\alpha_2|$ to become smaller, but, depending on the shape of the ellipsoid, $|\alpha_2|$ may not be much smaller than its limiting value:

$$-\alpha_2 \lesssim \eta_1. \tag{7.56}$$

This result is again in agreement with the experiments for PAA and MBBA. In the second model calculation, using the theory of dense fluids, Eq. (7.56) takes the form

$$-\alpha_2 = \frac{1-(b/a)^4}{1+(b/a)^2}\eta_1. \tag{7.57}$$

Furthermore, in that case one can derive the relation

$$\eta_2/\eta_1 = (b/a)^4. \tag{7.58}$$

Evidently this latter result is in serious disagreement with the data for PAA and MBBA (Tables 7.1 and 7.2). Incorporation of orientation fluctuations and molecular rotation in the model may be expected to lead to a comparatively larger value for η_2, which would be in better agreement with experiment.

Finally we come to a discussion of the temperature dependence of the viscosity coefficients. As is well known, the viscosity of an isotropic liquid varies approximately (see, for example, Frenkel, 1955) as

$$\eta_{\mathrm{is}} = \eta_0 \exp(E/k_{\mathrm{B}}T), \qquad (7.59)$$

where $E > 0$ is an activation energy for diffusion and η_0 is a constant. From Figure 7.4, we see that in the case of MBBA, the translational viscosities η_2 and η_3, to a first approximation, have the same temperature dependence as η_{is} ($E \approx 0.3$ eV). Close to T_{NI} there are deviations, however. As this is the region where S varies strongly, probably a more complicated formula depending also on S is involved. The temperature range, and thus the variation of S, is too limited to allow one to decide on the actual functional dependence, but the suggestion of Imura and Okano (1972) that

$$(\eta_2 - \eta_{\mathrm{is}})/S = \text{constant}, \qquad (7.60)$$

can be excluded. Here η_{is} is understood to be the isotropic viscosity extrapolated to the relevant temperature in the nematic phase. Figure 7.4 seems to indicate that the activation energy associated with η_1 is larger than that of η_{is}. As discussed in connection with Table 7.2, the results for η_1 of MBBA are probably not trustworthy because flow alignment cannot be excluded. Therefore we shall refrain from a further discussion of these data. We must conclude that there is a need for accurate measurements of the translational viscosities for compounds or mixtures with a wide nematic range. Provided that measurements of the order parameter are also available, this would allow a determination of the dependence of the viscosities on T and S. At present this is not possible.

With respect to the rotational viscosities, γ_1 and γ_2, somewhat more information is available. From the full line in Figure 7.10 and the data for the anisotropy of the magnetic susceptibility of MBBA, one finds that a plot of $\ln(\gamma_1/\Delta\chi^m)$ against $1/T$ gives a straight line. The same result has been found by Prost *et al.* (1976) for a mixture of the two isomers of *p*-methoxy-*p'*-butylazoxybenzene (N4, Merck, Darmstadt). This mixture has a rather wide nematic range and $\Delta\chi^m$ varies by a factor of three while γ_1 varies over more than a factor of thirty (see Figure 7.11). Assuming, as before, rotational symmetry around the long molecular axis, we have

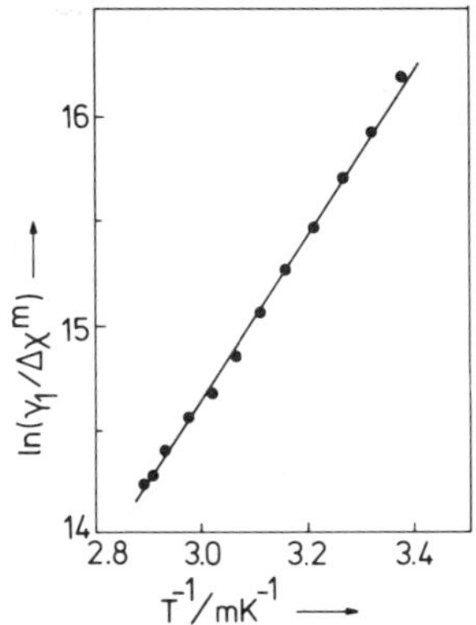

FIGURE 7.11   Rotational viscosity γ_1 *versus* $1/T$ for N4 (Prost *et al.* 1976).

$\Delta\chi^m \sim S$, leading to

$$\gamma_1 \sim S\exp(E'/k_B T). \tag{7.61}$$

Other functional dependences such as $\exp(E/k_B T)$, $S^2\exp(E/k_B T)$, $S^2\exp(ES/k_B T)$ and $S\exp(ES/k_B T)$ can clearly be ruled out. For MBBA one has $E' \approx 0.5$ eV (de Jeu, 1978b), which is rather higher than the activation energy associated with η_{is}. Information about the functional dependence of γ_2 can be obtained in the following way. Although γ_1 and γ_2 have no isotropic counterpart and vanish for $S=0$, there is still flow alignment of the molecules above T_{NI}. Two causes may contribute to this effect. Firstly, in any isotropic liquid consisting of elongated molecules, the molecules tend to align with their long axes at an angle of $45°$ with the flow and with the velocity gradient (see, for example, Frenkel, 1955). Secondly, above T_{NI} there is considerable pretransitional nematic order. Comparison of results from light scattering experiments above T_{NI} with viscosity data for the nematic phase leads to the conclusion that γ_2/γ_1 (which parameter governs the coupling of molecular orientation to shear strain) is continuous at T_{NI} (Clark, 1973). Hence it must be expected that γ_2 has the same functional dependence on S as γ_1.

Imura and Okano (1972) have given an extensive theoretical discussion of the temperature dependence of the viscosity coefficients of a nematic. They assume that the tensorial coefficients in the formal phenomenological equations that govern the hydrodynamic behaviour of nematics are to be constructed by the tensor order parameter, Eq. (6.1). In this way they derive Leslie's stress tensor, Eq. (7.27), with S-dependent viscosity coefficients. Up to the order S^2, these are given by

$$\alpha_1 = A_1 S^2,$$
$$\alpha_2 = -(B_1 + C_1)S - (B_2 + C_2)S^2,$$
$$\alpha_3 = -(B_1 - C_1)S - (B_2 - C_2)S^2,$$
$$\alpha_4 = 2\eta_{is} - aS + A_3 S^2, \tag{7.62}$$
$$\alpha_5 = \left(\tfrac{3}{2}a + B_1\right)S + (A_2 + B_2)S^2,$$
$$\alpha_6 = \left(\tfrac{3}{2}a - B_1\right)S + (A_2 - B_2)S^2.$$

The new coefficients a, B_i, C_i $(i = 1, 2)$ are constants which are expected to depend on temperature very weakly. The Onsager-Parodi relation is satisfied for each order of S. The only coefficient that is predicted to be of the order S^2 is α_1. Indeed, for MBBA, α_1 is observed to be much smaller than the other Leslie coefficients (see Table 7.2). The equations contain only one activation energy, which is common both to the nematic and the isotropic phase. Consequently, the temperature dependence of the rotational viscosities is predicted to depend only on that of the order parameter. From Eq. (7.62) we obtain

$$\gamma_1 = 2C_1 S + 2C_2 S^2,$$
$$-\gamma_2 = 2B_1 S + 2B_2 S^2, \tag{7.63}$$

in serious disagreement with the experiments discussed above. Eq. (7.63) was proposed independently by Helfrich (1972), who argued that $C_1 = 0$. This is because γ_1 must be positive, as discussed earlier, and S can be negative, at least theoretically. Following this argument, Martins and Diogo (1975) calculated C_2 from Maier and Saupe's molecular-statistical theory. This leads to

$$C_2 \sim \exp(ES/k_B T). \tag{7.64}$$

We must conclude that all these theoretical predictions are in rather poor agreement with the experimental results. Although the amount of experimental data at present available is limited, the experimentally established temperature dependence of γ_1, Eq. (7.61), requires

$$C_1 \sim \exp(E'/k_B T),$$
$$C_2 \approx 0. \tag{7.65}$$

We conclude that there is ample room for new theoretical approaches.

Many flow experiments on nematics have been performed in which the orientation of the director is not controlled (Porter and Johnson, 1966; Berchet *et al.* 1970). In that case, an effective viscosity is observed that depends on the flow alignment angle. Far from T_{NI} one often finds that θ_0 is small and approximately η_2 is measured. However, if T_{NI} is approached, θ_0 may increase (see Figure 7.6). Then a value in between η_2 and η_1 will be measured. An increase of θ_0 can explain the increase in the effective viscosity that is often observed near T_{NI}. It is clear that in the absence of further information on the state of alignment little can be said about such data.

We shall finish this chapter with a few remarks on the viscosities of smectic liquid crystals. The apparent viscosities of smectics, as measured by capillary methods, are very high. This is not due to large friction coefficients in the sense discussed above, but to an additional mechanism: permeation. Movement of the molecules normal to the layer occurs as through porous plugs which they themselves form. In addition the presence of defects will play a role. When a nematic-smectic A phase transition is approached, pretransitional smectic order may cause some of the viscosity coefficients to diverge, just like the twist and bend elastic constants. Such a pretransitional increase can be expected for α_1, α_3 and α_6, but not for α_2, α_4 and α_5 (Jähnig and Brochard, 1974; McMillan, 1974). Consequently, of the Miesowicz viscosities only η_2 will be affected, and in addition both γ_1 and γ_2. For γ_1 this effect has been investigated experimentally (Hardouin *et al.* 1974). We shall not discuss these effects here any further. However, it is clear from these comments that if one wants to probe the "regular" temperature dependence of the nematic viscosities, one should check that pretransitional smectic order is absent.

Some relations from vector and tensor calculus

$$a = (a_x, a_y, a_z) \quad \text{(vector)}$$

$$a \cdot b = a_x b_x + a_y b_y + a_z b_z \quad \text{(scalar)}$$

$$a \times b = (a_y b_z - a_z b_y, a_z b_x - a_x b_z, a_x b_y - a_y b_x) \quad \text{(vector)}$$

$$a \times a = 0$$

$$a \cdot (a \times b) = 0$$

$$a \cdot (b \times c) = (a \times b) \cdot c$$

$$a \times (b \times c) = b(a \cdot c) - c(a \cdot b)$$

$$\nabla = \left(\frac{\partial}{\partial x}, \frac{\partial}{\partial y}, \frac{\partial}{\partial z} \right) \quad \text{(vector operator)}$$

$$\nabla \phi = \operatorname{grad} \phi = \left(\frac{\partial \phi}{\partial x}, \frac{\partial \phi}{\partial y}, \frac{\partial \phi}{\partial z} \right) \quad \text{(vector)}$$

$$\nabla \cdot a = \operatorname{div} a = \frac{\partial a_x}{\partial x} + \frac{\partial a_y}{\partial y} + \frac{\partial a_z}{\partial z} \quad \text{(scalar)}$$

$$\nabla \times a = \operatorname{rot} a = \left(\frac{\partial a_z}{\partial y} - \frac{\partial a_y}{\partial z}, \frac{\partial a_x}{\partial z} - \frac{\partial a_z}{\partial x}, \frac{\partial a_y}{\partial x} - \frac{\partial a_x}{\partial y} \right) \quad \text{(vector)}$$

$$\nabla^2 = \frac{\partial^2}{\partial x^2} + \frac{\partial^2}{\partial y^2} + \frac{\partial^2}{\partial z^2} \quad \text{(scalar operator)}$$

$$\nabla^2 a = \left(\frac{\partial^2 a_x}{\partial x^2}, \frac{\partial^2 a_y}{\partial y^2}, \frac{\partial^2 a_z}{\partial z^2} \right) \quad \text{(vector)}$$

Gauss' theorem:

$$\int_V \nabla \cdot a \, dV = \int_A a \cdot \sigma \, dA,$$

where V is the volume inside the closed surface A; σ is a unit vector normal to the surface.

Stokes' theorem:

$$\oint_L a \cdot dl = \int_A (\nabla \times a) \cdot \sigma \, dA,$$

where A is any surface bounded by the contour L.

A tensor of second rank $\mathbf{T}$ describes the linear relation between two vectors that do not have the same direction:

$$a = \mathbf{T} \cdot b,$$

or in components

$$a_\alpha = (\mathbf{T} \cdot b)_\alpha = \sum_\beta T_{\alpha\beta} b_\beta, \qquad \alpha, \beta = x, y, z.$$

In a matrix representation this can be written as

$$\begin{pmatrix} a_1 \\ a_2 \\ a_3 \end{pmatrix} = \begin{pmatrix} T_{11} & T_{12} & T_{13} \\ T_{21} & T_{22} & T_{23} \\ T_{31} & T_{32} & T_{33} \end{pmatrix} \begin{pmatrix} b_1 \\ b_2 \\ b_3 \end{pmatrix}.$$

Any *symmetric* tensor of rank two (for which $T_{\alpha\beta} = T_{\beta\alpha}$) can be put in diagonal form by choosing a suitable set of coordinate axis k, l, m (principal axes):

$$\mathbf{T} = \begin{pmatrix} T_{kk} & 0 & 0 \\ 0 & T_{ll} & 0 \\ 0 & 0 & T_{mm} \end{pmatrix}.$$

The trace of the matrix representing a tensor

$$\mathrm{Tr}(\mathbf{T}) = \sum_\alpha T_{\alpha\alpha},$$

is invariant under such a transformation.

In conclusion we can form the following cross products of two vectors

$$a \cdot b = \sum_\alpha a_\alpha b_\alpha, \qquad \text{(scalar)}$$

$$ab = T \quad \text{with} \quad T_{\alpha\beta} = a_\alpha b_\beta. \qquad \text{(tensor)}$$

In the same way we have for two tensors $\mathbf{T}$ and $\mathbf{S}$

$$(\mathbf{T} \cdot \mathbf{S})_{\alpha\beta} = \sum_\gamma T_{\alpha\gamma} S_{\gamma\beta} \qquad \text{(tensor)}$$

$$\mathbf{T} : \mathbf{S} = \sum_{\alpha, \beta} T_{\alpha\beta} S_{\beta\alpha} \qquad \text{(scalar)}.$$

References

General references

G. H. Brown, ed., *Advances in Liquid Crystals*, Academic Press, New York; vol. I, 1975; vol. II, 1976; vol. III, 1978; vol. IV, 1979.

S. Chandrasekhar, *Liquid Crystals*, Cambridge University Press, Cambridge, 1977.

D. Demus, H. Demus, and H. Zaschke, *Flüssige Kristalle in Tabellen*, VEB Verlag, Leipzig, 2nd. ed., 1976.

D. Demus and L. Richter, *Textures of Liquid Crystals*, VEB Verlag, Leipzig, 1978.

P. G. de Gennes, *The Physics of Liquid Crystals*, Clarendon Press, Oxford, 1974.

G. W. Gray and P. A. Winsor, eds., *Liquid Crystals and Plastic Crystals*, Wiley, New York, 1974, vols. I and II.

H. Kelker and R. Hatz, *Handbook of Liquid Crystals*, Verlag Chemie, Weinheim, 1979.

L. Liebert, ed., *Liquid Crystals*, Solid State Physics Suppl. no. 14, Academic Press, New York, 1978.

G. R. Luckhurst and G. W. Gray, eds., *The Molecular Physics of Liquid Crystals*, Academic Press, London, 1979.

E. B. Priestley, P. J. Wojtowicz, and P. Sheng, eds., *Introduction to Liquid Crystals*, Plenum, New York, 1974.

M. J. Stephen and J. P. Straley, Rev. Mod. Phys. **46**, 617 (1974).

See further the journal Molecular Crystals and Liquid Crystals, and the special colloquium issues of the Journal de Physique: C-1 (1975), C-3 (1976), C-3 (1979).

References cited in text

V. K. Agarwal and A. H. Price (1974), JCS Faraday Trans. II, **70**, 188.

R. Alben, J. R. McColl, and C. S. Shih (1972), Solid State Commun. **11**, 1081.

R. S. Armstrong and R. J. W. Le Fèvre (1966), Aust. J. Chem. **19**, 29.

H. Arnold (1964), Z. Phys. Chem. (Leipzig) **226**, 146.

A. Axmann (1968), Mol. Cryst. **3**, 471.

J. C. Bacri (1974), J. Phys. Lettres **35**, L-141.

C. C. Ballard, E. C. Broge, R. K. Iler, D. S. St. John, and J. R. McWhorther (1961), J. Phys. Chem. **65**, 20.

L. Bata and G. Molnar (1975), Chem. Phys. Lett. **33**, 535.

D. Berchet, A. Hochapfel, and R. Viovy (1970), C. R. Hebd. Séan. Acad. Sci. **C270**, 1065.

D. W. Berreman (1972), Phys. Rev. Lett. **28**, 1683.

J. Billard, J. C. Dubois, Nguyen Huu Tinh, and A. Zann (1978), Nouv. J. Chim. **2**, 535.

L. M. Blinov, V. A. Kizel, V. G. Rumyantsev, and V. V. Titov (1975), J. Phys. **36**, C1-69.

H.-P. Boehm (1966), Angew. Chem. **78**, 617.

P. Bordewijk (1973), Physica **69**, 422.

P. Bordewijk (1974), Physica **75**, 146.

P. Bordewijk and W. H. de Jeu (1978), J. Chem. Phys. **68**, 116.

C. J. F. Böttcher (1973), *Theory of Electric Polarization*, 2nd. ed., Elsevier (Amsterdam), vol. I.

C. J. F. Böttcher and P. Bordewijk (1978), *Theory of Electric Polarization*, 2nd. ed., Elsevier (Amsterdam), vol. II.

F. Brochard (1972), J. Phys. **33**, 607.

M. Brunet-Germain (1970), C. R. Hebd. Séan. Acad. Sci. **B271**, 1075.

C. H. Carlisle and C. H. Smith (1971), Acta Crystallogr. **B27**, 1068.
S. Chandrasekhar and N. V. Madhusudana (1969), J. Phys. **30**, C4-24.
S. Chandrasekhar, B. K. Sadashiva, and K. A. Suresh (1977), Pramana **9**, 471.
P. Chatelain (1936), C. R. Hebd. Séan. Acad. Sci. **203**, 1169.
P. Chatelain and M. Germain (1964), C. R. Hebd. Séan. Acad. Sci. **259**, 127.
L. Cheung, R. B. Meyer, and H. Gruler (1973), Phys. Rev. Lett. **31**, 349.
P. E. Cladis (1972), Phys. Rev. Lett. **28**, 1629.
P. E. Cladis (1973), Phys. Rev. Lett. **31**, 1200.
P. E. Cladis, J. Rault and J.-P. Burger (1971), Mol. Cryst. Liq. Cryst. **13**, 1.
P. E. Cladis and S. Torza (1975), Phys. Rev. Lett. **35**, 1283.
N. A. Clark (1973), Phys. Lett. **46A**, 171.
M. Davies, R. Moutran, A. H. Price, M. S. Beevers, and G. Williams (1976), JSC Faraday II, **72**, 1447.
M. Delaye, R. Ribotta, and G. Durand (1973), Phys. Rev. Lett. **31**, 443.
A. I. Derzhanski and A. G. Petrov (1971), C. R. Acad. Bulg. Sci. **24**, 569, 573.
H. Deuling (1978), p. 77 in *Liquid Crystals*, L. Liebert ed., Solid State Physics Suppl. no. 14, Academic Press (New York).
P. Diehl and C. L. Khetrapal (1969), p. 1 in *NMR, Basic Principles and Progress*, P. Diehl, E. Fluck and R. Kosfeld, eds., vol. I, Springer (Berlin).
D. Diguet, F. Rondelez, and G. Durand (1970), C. R. Hebd. Séan. Acad. Sci. **B271**, 954.
D. Dolphin, Z. Muljiani, J. Cheng, and R. B. Meyer (1973), J. Chem. Phys. **58**, 413.
J. Doucet, A. M. Levelut, and M. Lambert (1974), Phys. Rev. Lett. **32**, 301.
C. Druon and J. M. Wacrenier (1977), J. Phys. **38**, 47.
E. Dubois-Violette, G. Durand, E. Guyon, P. Manneville, and P. Pieranski (1978), p. 147 in *Liquid Crystals*, L. Liebert, ed., Solid State Physics Suppl. no. 14, Academic Press (New York).
D. A. Dunmur and W. H. Miller (1979), J. Phys. Colloque **40**, C3-141.
G. Durand, L. Leger, F. Rondelez, and M. Veyssie (1969), Phys. Rev. Lett. **22**, 227.
M. Dvolaitzky, J. Billard, and F. Poldy (1976), Tetrahedron **32**, 1835.
D. C. van Eck and W. Westera (1977), Mol. Cryst. Liq. Cryst. **38**, 319.
D. C. van Eck and R. J. J. Zijlstra (1978), p. 270 in *Noise in Physical Systems*, D. Wolff, ed., Springer (Berlin).
R. Eidenschink, D. Erdmann, J. Krause and L. Pohl (1978), Angew. Chem. **90**, 133.
W. Elser and R. D. Ennulat (1976), p. 73 in *Advances in Liquid Crystals*, G. H. Brown, ed., vol. II, Academic Press (New York).
J. W. Emsley, J. C. Lindon, and G. R. Luckhurst (1975), Mol. Phys. **30**, 1913.
J. W. Emsley, G. R. Luckhurst, G. W. Gray, and A. Mosley (1978), Mol. Phys. **35**, 1499.
J. L. Ericksen (1976), p. 233 in *Advances in Liquid Crystals*, G. H. Brown, ed., vol. II, Academic Press (New York).
W. H. Flygare (1974), Chem. Rev. **74**, 653.
G. Föex (1933), Trans. Faraday Soc. **29**, 958.
F. C. Frank (1958), Disc. Faraday Soc. **25**, 19.
J. Frenkel (1955), *Kinetic Theory of Liquids*, Dover (New York).
Ch. Gähwiller (1972), Phys. Rev. Lett. **28**, 1554.
Ch. Gähwiller (1973), Mol. Cryst. Liq. Cryst. **20**, 301.
H. Gasparoux and J. Prost (1971), J. Phys. **32**, 953.
H. Gasparoux, B. Regaya, and J. Prost (1971), C. R. Hebd. Séan. Acad. Sci. **B272**, 1168.
P. G. de Gennes (1968), C. R. Hebd. Séan. Acad. Sci. **B266**, 15.
P. G. de Gennes (1971), Mol. Cryst. Liq. Cryst. **12**, 193.
W. F. Gorham (1966), J. Polym. Sci. part A-1, **4**, 3027.
G. W. Gray (1975), J. Phys. Colloque **36**, C1-337.
H. Gruler (1974), J. Chem. Phys. **61**, 5408.
H. Gruler (1975), Z. Naturforsch. **30a**, 230.
H. Gruler and L. Cheung (1975), J. Appl. Phys. **46**, 5097.
H. Gruler, T. J. Scheffer, and G. Meier (1972), Z. Naturforsch. **27a**, 966.
E. A. Guggenheim (1967), *Thermodynamics*, 5th ed., North-Holland (Amsterdam).

E. Guyon and W. Urbach (1976), p. 121 in *Nonemissive electrooptic displays*, A. R. Kmetz and F. K. von Willisen, eds., Plenum (New York).

W. Haberdizl (1976), Sec. 3.5E in *Theory and Applications of Molecular Diamagnetism*, L. N. Mulay and E. A. Boudreaux, eds., Wiley (New York).

I. Haller (1972), J. Chem. Phys. **57**, 1400.

I. Haller, H. A. Huggins, and M. J. Freiser (1972), Mol. Cryst. Liq. Cryst. **16**, 53.

I. Haller, H. A. Huggins, H. R. Lilienthal, and T. R. McGuire (1973), J. Phys. Chem. **77**, 950.

E. G. Hanson and Y. R. Shen (1976), Mol. Cryst. Liq. Cryst. **36**, 193.

F. Hardouin, M. F. Achard, and H. Gasparoux (1974), Solid State Commun. **14**, 453.

W. Helfrich (1969), J. Chem. Phys. **50**, 100.

W. Helfrich (1970), J. Chem. Phys. **53**, 2267.

W. Helfrich (1972), J. Chem. Phys. **56**, 3187.

G. Heppke and F. Schneider (1972), Z. Naturforsch **27a**, 976.

N. E. Hill (1969), Ch. 1 in *Dielectric properties and Molecular Behaviour*, N. E. Hill, W. E. Vaughan, A. H. Price, and M. Davies, Van Nostrand Reinhold (London).

A. Höhener, L. Müller, and R. R. Ernst (1979), Mol. Phys. **38**, 909.

R. G. Horn (1978a), J. Phys. **39**, 105.

R. G. Horn (1978b), J. Phys. **39**, 167.

H. Imura and K. Okano (1972), Jap. J. Appl. Phys. **11**, 1440.

F. Jähnig and F. Brochard (1974), J. Phys. **35**, 301.

S. Jen, N. A. Clark, P. S. Pershan, and E. B. Priestley (1977), J. Chem. Phys. **66**, 4635.

W. H. de Jeu (1977), J. Phys. **38**, 1265.

W. H. de Jeu (1978a), p. 109 in *Liquid Crystals*, L. Liebert, ed., Solid State Physics Suppl. no. 14, Academic Press (New York).

W. H. de Jeu (1978b), Phys. Lett. **69A**, 122.

W. H. de Jeu and P. Bordewijk (1978), J. Chem. Phys. **68**, 109.

W. H. de Jeu and W. A. P. Claassen (1977), J. Chem. Phys. **67**, 3705.

W. H. de Jeu and W. A. P. Claassen (1978), J. Chem. Phys. **68**, 102.

W. H. de Jeu, W. A. P. Claassen, and A. M. J. Spruijt (1976), Mol. Cryst. Liq. Cryst. **37**, 269.

W. H. de Jeu, C. J. Gerritsma, P. van Zanten, and W. J. A. Goossens (1972), Phys. Lett. **39A**, 355.

W. H. de Jeu, W. J. A. Goossens, and P. Bordewijk (1974), J. Chem. Phys. **61**, 1985.

W. H. de Jeu and T. W. Lathouwers (1974), Mol. Cryst. Liq. Cryst. **26**, 225.

W. H. de Jeu and T. W. Lathouwers (1975), Z. Naturforsch. **30a**, 79.

W. H. de Jeu and F. Leenhouts (1978), J. Phys. **39**, 869.

W. H. de Jeu and J. van der Veen (1973), Phys. Lett. **44A**, 277.

W. H. de Jeu and J. van der Veen (1977), Mol. Cryst. Liq. Cryst. **40**, 1.

F. J. Kahn, G. N. Taylor, and H. Schonhorn (1973), Proc. IEEE, **61**, 823.

H. Kelker (1973), Mol. Cryst. Liq. Cryst. **21**, 1.

H. Kelker, B. Scheurle, R. Hatz, and W. Bartsch (1970), Angew. Chem. **82**, 984.

P. Keller and L. Liebert (1978), p. 20 in *Liquid Crystals*, L. Liebert, ed., Solid State Physics Suppl. no. 14, Academic Press (New York).

M. G. Kim, S. Park, Sr. M. Cooper, and S. V. Letcher (1976), Mol. Cryst. Liq. Cryst. **36**, 143.

K. Kiry and P. Martinoty (1977), J. Phys. Lettres **38**, L-389.

R. T. Klingbiel, D. J. Genova, and H. K. Bücher (1974a), Mol. Cryst. Liq. Cryst. **27**, 1.

R. T. Klingbiel, D. J. Genova, T. R. Criswell, and J. P. van Meter (1974b), J. Am. Chem. Soc. **96**, 7651.

H. Kresse, P. Schmidt, and D. Demus (1975), Phys. Stat. Sol. **a32**, 315.

H. Kresse, K.-H. Lücke, P. Schmidt, and D. Demus (1977), Wiss. Z. Univ. Halle **26**, 147.

W. R. Krigbaum, Y. Chatani, and P. G. Barber (1970), Acta Crystallogr. **B26**, 97.

A. N. Kuznetsov, V. A. Livshits, and S. G. Cheskis (1975), Sov. Phys. Crystallogr. **20**, 142.

G. Labrunie and M. Bresse (1973), C. R. Hebd. Séan. Acad. Sci. **B276**, 647.

L. D. Landau and E. M. Lifshitz (1959), *Fluid Mechanics*, Pergamon (Oxford).

A. J. Leadbetter, J. L. A. Durrant, and M. Rugman (1977), Mol. Cryst. Liq. Cryst. Letters **34**, 231.

L. B. Leder (1971), J. Chem. Phys. **55**, 2649.

F. Leenhouts, A. J. Dekker, and W. H. de Jeu (1979a), Phys. Lett. **72A**, 155.
F. Leenhouts and F. van der Woude (1978), J. Phys. Lettres **39**, L-249.
F. Leenhouts, F. van der Woude, and A. J. Dekker (1976), Phys. Lett. **58A**, 242.
F. Leenhouts, H. J. Roebers, A. J. Dekker, and J. J. Jonker (1979b), J. Phys. Colloque **40**, C3-291.
R. J. W. Le Fèvre (1965), p. 1 in *Adv. Phys. Org. Chem.*, V. Gold, ed., vol. 3, Academic Press (London).
L. Léger (1972), Solid State Commun. **11**, 1499.
F. M. Leslie (1970), Mol. Cryst. Liq. Cryst. **12**, 57.
F. M. Leslie (1979), p. 1 in *Advances in Liquid Crystals*, G. H. Brown, ed., vol. IV, Academic Press (New York).
D. Lippens, J. P. Parneix, and A. Chapoton (1977), J. Phys. **38**, 1465.
G. R. Luckhurst and C. Zannoni (1975), Proc. Roy. Soc. **A343**, 389.
G. R. Luckhurst, C. Zannoni, P. L. Nordio, and U. Segre (1975), Mol. Phys. **30**, 1345.
N. Lumbroso-Bader (1956), Ann. Chim. **1**, 687.
N. V. Madhusudana, P. P. Karat, and S. Chandrasekhar (1975), p. 225 in *Proc. Bangalore Liq. Cryst. Conf.*, Pramana, suppl. no. 1.
W. Maier and G. Meier (1961a), Z. Naturforsch. **16a**, 262.
W. Maier and G. Meier (1961b), Z. Naturforsch. **16a**, 470.
W. Maier and G. Meier (1961c), Z. Elektrochem. **65**, 556.
W. Maier and G. Meier (1961d), Z. Naturforsch. **16a**, 1200.
W. Maier and A. Saupe (1959), Z. Naturforsch. **14a**, 882.
W. Maier and A. Saupe (1960), Z. Naturforsch. **15a**, 287.
H. Margenau and G. M. Murphy (1956), *The Mathematics of Physics and Chemistry*, 2nd. ed., Van Nostrand (Princeton).
V. Marinin and V. Tsvetkov (1939), Acta Physicochim. USSR **11**, 837.
A. J. Martin, G. Meier, and A. Saupe (1971), Symp. Faraday Soc. **5**, 119.
J. L. Martinand and G. Durand (1972), Solid State Commun. **10**, 815.
P. Martinoty and S. Candau (1971), Mol. Cryst. Liq. Cryst. **14**, 243.
A. F. Martins and A. C. Diogo (1975), Portugal. Phys. **9**, 129.
S. Matsumoto, K. Mizunoya, D. Nakagawa, N. Kaneko, and M. Kawamoto (1979), J. Phys. Colloque **40**, C3-510.
W. L. McMillan (1974), Phys. Rev. **A9**, 1720.
S. Meiboom and R. C. Hewitt (1973), Phys. Rev. Lett. **30**, 261.
G. Meier and A. Saupe (1966), Mol. Cryst. **1**, 515.
M. Miesowicz (1936), Bull. Intern. Acad. Polon. Ser. A, 228.
V. I. Minkin, D. A. Osipov and Yu. A. Zhdanov (1970), *Dipole Moments in Organic Chemistry*, Plenum (New York).
A. Mircéa-Roussel and F. Rondelez (1975), J. Chem. Phys. **63**, 2311.
L. N. Mulay (1976), Sec. 5 in *Theory and Applications of Molecular Diamagnetism*, L. N. Mulay and E. A. Boudreaux, eds., Wiley (New York).
J. Nehring and A. Saupe (1971), J. Chem. Phys. **54**, 337.
J. Nehring and A. Saupe (1972), J. Chem. Phys. **56**, 5527.
H. E. J. Neugebauer (1954), Can J. Phys. **32**, 1.
C. S. Oh (1977), Mol. Cryst. Liq. Cryst. **42**, 1.
Orsay Liquid Crystal Group (1969), J. Chem. Phys. **51**, 816.
Orsay Liquid Crystal Group (1971), Mol. Cryst. Liq. Cryst. **13**, 187.
A. Pacault, J. Hoarau, and J. Favède (1967), p. 141 in Landolt-Börnstein, vol. 10, Springer (Berlin).
J. P. Parneix, A. Chapoton, and E. Constant (1975), J. Phys. **36**, 1143.
O. Parodi (1970), J. Phys. **31**, 581.
R. S. Porter and J. F. Johnson (1966), J. Chem. Phys. **45**, 1452.
G. Pelzl, R. Rettig, and D. Demus (1975), Z. Phys. Chem. (Leipzig) **256**, 305.
G. Pelzl and H. Sackmann (1971), Symp. Faraday Soc. **5**, 68.
G. Pelzl and H. Sackmann (1973), Z. Phys. Chem. (Leipzig) **254**, 354.
P. Pieranski, F. Brochard, and E. Guyon (1973), J. Phys. **34**, 35.

P. Pieranski and E. Guyon (1974), Phys. Rev. Lett. **32**, 924.
A. Pines and J. J. Chang (1974), Phys. Rev. **A10**, 946.
A. Pines and A. Höhener (1976), private communication.
L. Pohl, R. Eidenschink, J. Krause, and G. Weber (1978), Phys. Lett. **65A**, 169.
E. J. Poziomek, T. J. Novak, and R. A. Mackay (1972), Mol. Cryst. Liq. Cryst. **15**, 283.
R. G. Priest (1972), Mol. Cryst. Liq. Cryst. **17**, 129.
R. G. Priest (1973), Phys. Rev. **A7**, 720.
J. Prost, S. Sigaud, and B. Regaya (1976), J. Phys. Lettres **37**, L-341.
J. E. Proust and L. Ter-Minassian-Saraga (1975), J. Phys. **36**, C1-77.
B. Regaya and H. Gasparoux (1971), C. R. Hebd. Séan. Acad. Sci. **B272**, 724.
F. Rondelez and A. Mircea-Roussel (1974), Mol. Cryst. Liq. Cryst. **28**, 173.
J. C. Rowell, W. D. Phillips, L. R. Melby, and M. Panar (1965), J. Chem. Phys. **43**, 3442.
E. I. Ryumtsev, A. P. Kovshik, I. P. Kolomiets, and V. N. Tsvetkov (1974), Sov. Phys. Crystallogr. **18**, 780.
M. V. Sargent and C. J. Timmons (1964), J. Chem. Soc., 5544.
A. Saupe (1960), Z. Naturforsch. **15a**, 815.
A. Saupe (1963), Z. Naturforsch. **19a**, 161.
A. Saupe (1973), Ann. Rev. Phys. Chem. **24**, 441.
A. Saupe and W. Maier (1961), Z. Naturforsch. **16a**, 816.
M. Schadt (1972), J. Chem. Phys. **56**, 1494.
M. Schadt and F. Müller (1979), Revue Phys. Appl. **14**, 265.
T. J. Scheffer and J. Nehring (1977), J. Appl. Phys. **48**, 1783.
H. Schubert and H. Dehne (1972), Z. Chem. **12**, 241.
H. Schulze and W. Burkersrode (1975), Exp. Techn. Phys. **23**, 369.
S. A. Shaya and H. Yu (1975), J. Chem. Phys. **63**, 221.
J. P. Straley (1973), Phys. Rev. **A8**, 2181.
H. Sorkin and A. Denny (1973), RCA Rev. **34**, 309.
Suh-Bong Rhee and H. H. Jaffé (1973), J. Am. Chem. Soc. **95**, 5518.
J. W. Summerford, J. R. Boyd, and B. A. Lowry (1975), J. Appl. Phys. **46**, 970.
H. C. Tseng, D. L. Silver, and B. A. Finlayson (1972), Phys. Fluids **15**, 1213.
V. N. Tsvetkov and G. M. Michailov (1938), Acta Physicochim. USSR **8**, 77.
V. N. Tsvetkov and A. Sosnovsky (1943), Acta Physicochim. USSR **18**, 358.
W. E. Vaughan (1969), Ch. 2 in *Dielectric Properties and Molecular Behaviour*, N. E. Hill, W. E. Vaughan, A. H. Price, and M. Davies, Van Nostrand Reinhold (London).
J. van der Veen and T. C. J. M. Hegge (1974), Angew. Chem. **86**, 378.
A. de Vries (1975), p. 93 in *Proc. Bangalore Liq. Cryst. Conf.*, Pramana, suppl. no. 1.
M. F. Vuks (1966), Optics Spectr. **20**, 361.
A. Weiss and H. Witte (1973), *Magnetochemie*, Verlag Chemie (Weinheim).
A. E. White, P. E. Cladis, and S. Torza (1977), Mol. Cryst. Liq. Cryst. **43**, 13.

List of Tables